ROB MALPELI
AMANDA TELFORD
CLAIRE STONEHOUSE
LEE ANTON-HEM
DEAN DUDLEY
SAM WATKINS
EMMÉ WILD
DAVID BAKKER

Nelson Fit For Life Health and Physical Education for The Australian Curriculum Levels 7 and 8 Workbook
2nd Edition
Rob Malpeli
Amanda Telford
Claire Stonehouse
Lee Anton-Hem
Dean Dudley
Sam Watkins
Emmé Wild
David Bakker

Product manager: Sarah Craig/Cathy Beswick-Davison
Content developer: Rachael Pictor
Project editor: Sutha Surenddar
Editor: Anne Mulvaney/MPS
Proofreader: Anne Mulvaney/MPS
Production controller: Karen Young/Sutha Surenddar
Permissions/Photo researcher: Jes Senbergs
Cover designer: Mariana Maccarini
Text designer: Mariana Maccarini
Project designer: Mariana Maccarini
Typeset by: MPS Limited
Cover: Shutterstock.com/Nicetoseeya
MPS Limited

Acknowledgements
We respectfully refer to Aboriginal and Torres Strait Islander Peoples as First Nations Peoples throughout the book.

ACKNOWLEDGEMENT OF COUNTRY
Nelson acknowledges the Traditional Owners and Custodians of the lands of all First Nations Peoples of Australia. We pay respect to their Elders past and present.
We recognise the continuing connection of First Nations Peoples to the land, air and waters, and thank them for protecting these lands, waters and ecosystems since time immemorial.

WARNING:
First Nations Peoples are advised that this book and associated learning materials may contain images, videos or voices of deceased persons.

For product information and technology assistance,
in Australia call 1300 790 853;
in New Zealand call 0800 449 725

For permission to use material from this text or product, please email aust.permissions@cengage.com

National Library of Australia Cataloguing-in-Publication Data
A catalogue record for this book is available from the National Library of Australia.
9780170465533

Cengage Learning Australia
Level 5, 80 Dorcas Street
Southbank VIC 3006 Australia

Cengage Learning New Zealand
Unit 4B Rosedale Office Park
331 Rosedale Road, Albany, North Shore 0632, NZ

For learning solutions, visit cengage.com.au

Printed in China by 1010 Printing International Limited.
3 4 5 6 7 26 25 24

CONTENTS

9780170465533

Chapter 9 Enhancing personal fitness through lifelong physical activity 181

Chapter 10 Just dance! 223

1

GET SMART ABOUT DRUGS

WORKSHEET 1.1 WHAT DO YOU KNOW ALREADY?

Page 4

Before we begin, let's see how much you already know about alcohol and other drugs. Complete each of the 'brainstorms' below with a list of words that you think of when you hear the words 'drugs' and 'alcohol'.

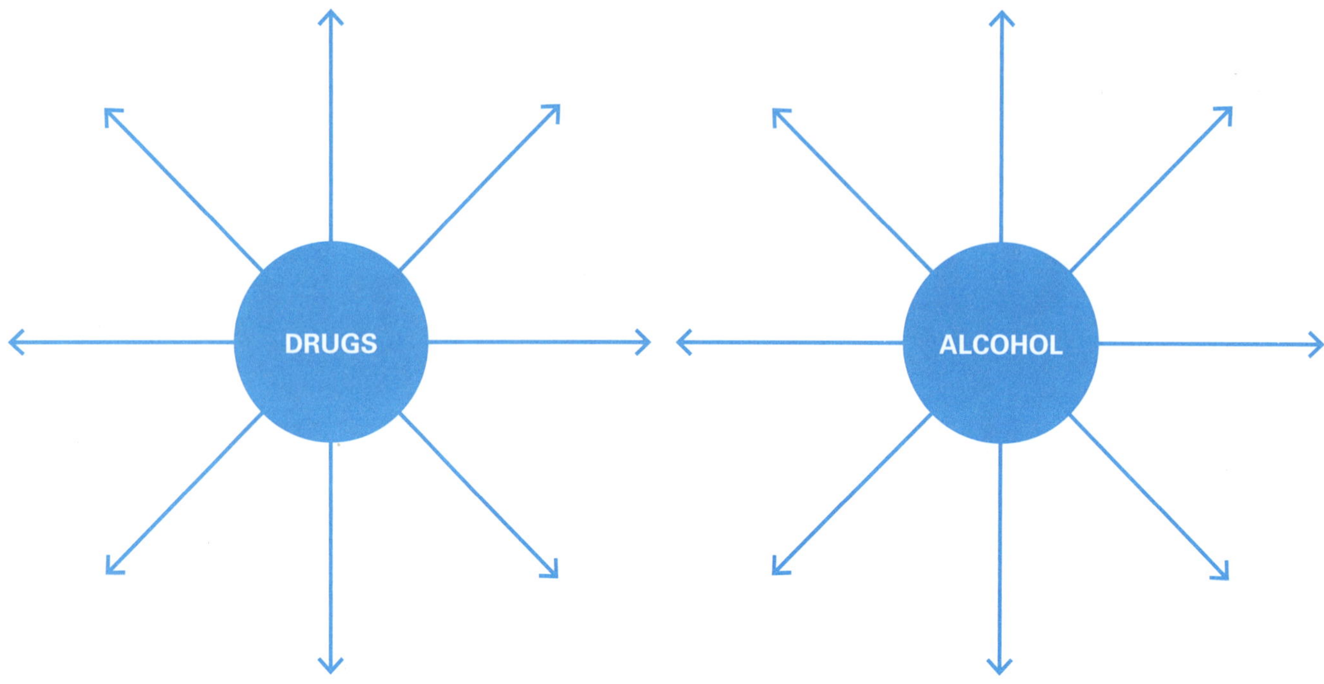

1 Think, pair, share

Compare to display recognition of similarities and differences and recognise the significance of these similarities and differences

Compare your lists of words with another student's lists. What words have you used that are similar and what words are different? Write a sentence that could be used as a definition for the terms 'drugs' and 'alcohol'.

Drugs are:

____

____

____

____

Alcohol is:

____

____

____

____

WORKSHEET 1.2 AGREE OR DISAGREE?

How strongly do you agree with the statements below? For each statement, place its number where you feel it would fit on the continuum below. There are no right or wrong answers.

1 Taking medicine is not harmful.

2 Increasing the price of alcohol will reduce over-consumption.

3 Vaping should be banned in all public places, both indoors and outdoors.

4 Marijuana should be legalised nationally.

5 Increasing penalties for drug use will stop people from taking drugs.

6 Energy drinks are potentially lethal.

Strongly disagree	Disagree	Neither agree nor disagree	Agree	Strongly agree

Justify where you placed each of the statements.

justify to show how an argument or conclusion is right or reasonable

AC

1 ______________________________

____

____

____

____

____

2 ______________________________

____

____

____

____

____

WORKSHEET 1.2 CONTINUED

3

4

5

6

WORKSHEET 1.3 PRESCRIPTION DRUG ABUSE

Pages 8–9

Did you know that Australia has one of the world's highest rates of prescription drug abuse?

After reading the 'How can I use medicine safely' section in your student book, fill out the PMI (Plus, Minus and Interesting) table below by identifying from the information what is a plus, what is a minus and what you found interesting.

PMI table on the safe use of medicine

Plus	Minus	Interesting

1 **Explain** the difference between over-the-counter (OTC) medicine and prescription medicine.

Explain to make an idea or situation plain or clear by describing it in more detail or revealing relevant facts

2 **Identify** two common medicines that can be bought over the counter.

Identify to provide an answer from a number of possibilities

WORKSHEET 1.3 CONTINUED

3 Identify two medicines that require a prescription.

consider to think deliberately or carefully about something, typically before making a decision

AC

4 **Consider** reasons why people purchase over-the-counter medication.

5 Medicine can be taken in a variety of ways. Name three different forms of prescribed and over-the-counter medications and provide an example of each.

discuss to talk or write about a topic, taking into account different issues or ideas

AC

6 **Discuss** reasons why people would choose to abuse either over-the-counter or prescription medication.

WORKSHEET 1.4 CULTURAL MEDICINES

Traditional medicine dates back thousands of years and is still used by many cultures throughout the world to promote healing and maintain health and wellbeing. Conduct an online search of the following civilisations and the types of traditional medicines that they used. On each scroll, write a brief history of the type of medicine, draw a diagram, and illustrate the way the medicine was used.

SB Pages 10–13

Indigenous Australian bush medicine

Ancient Egyptian medicine

Chinese medicine

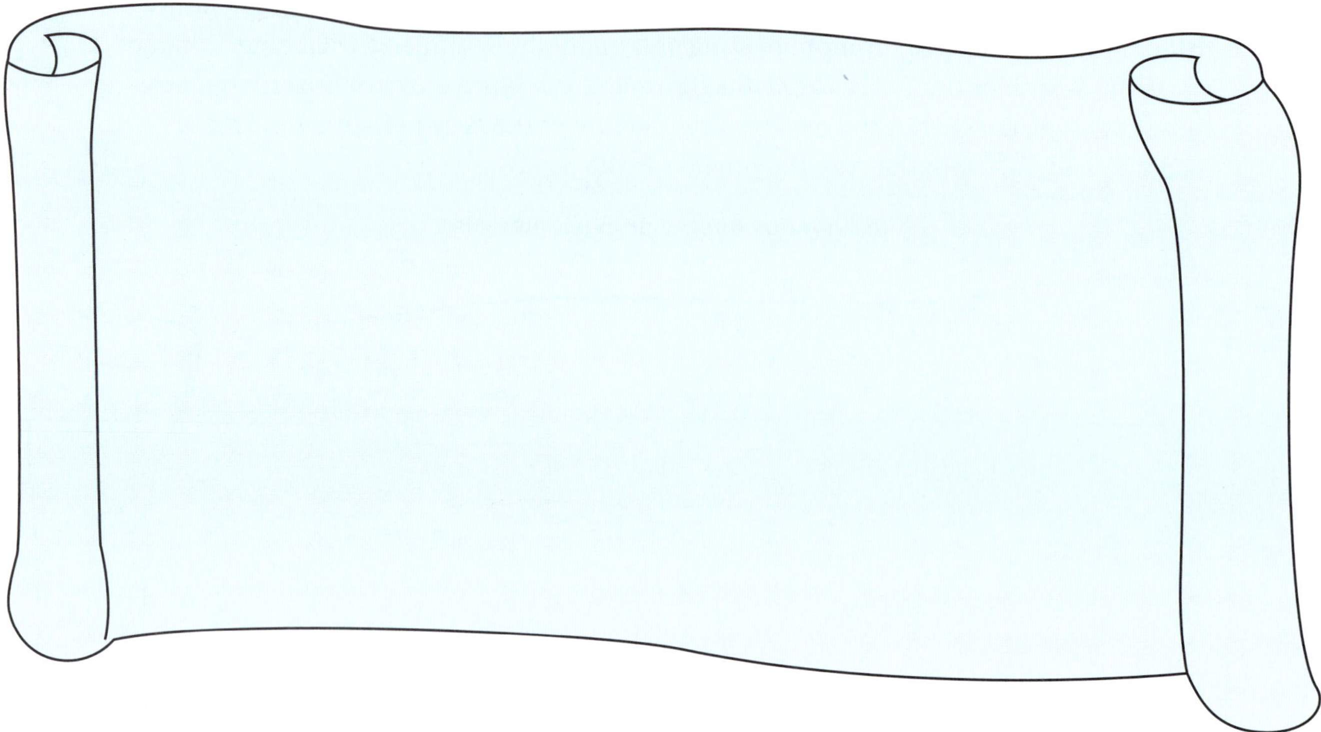

Ayurvedic medicine

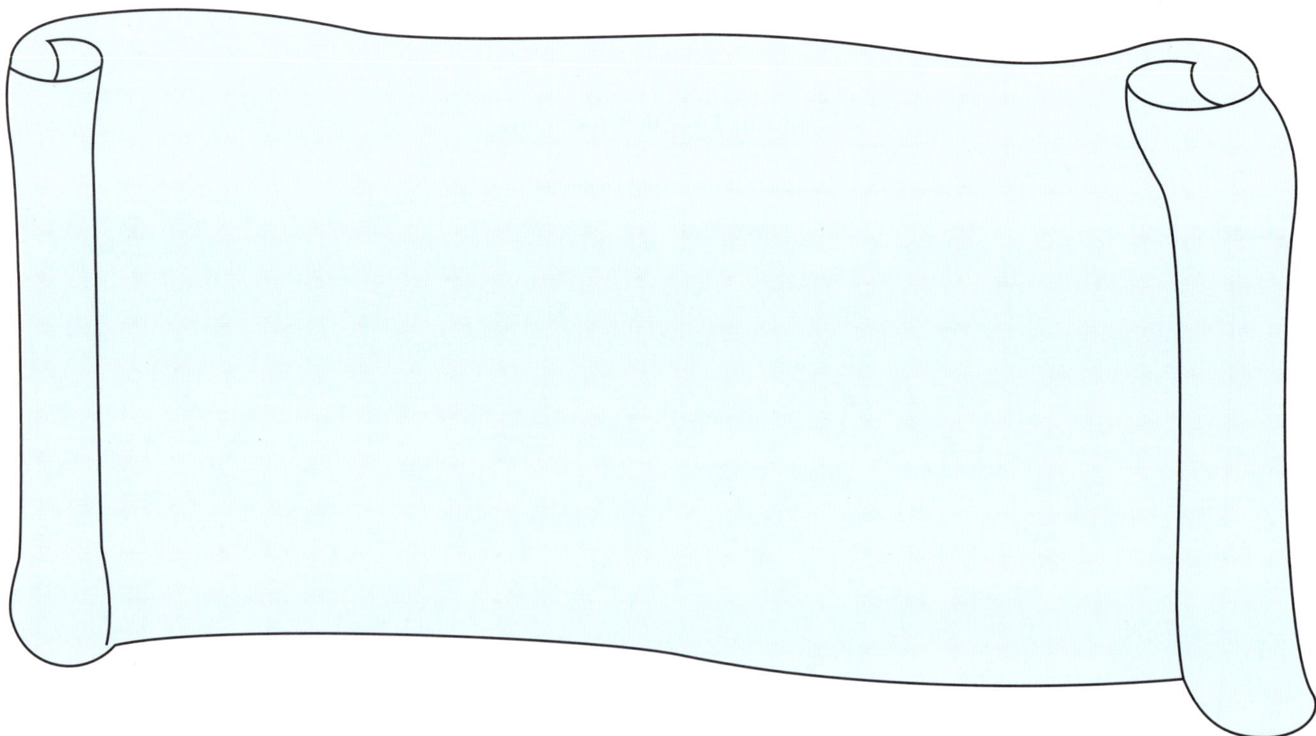

WORKSHEET 1.5 AUSTRALIA'S PERCEIVED PROBLEM WITH ALCOHOL

The Foundation for Alcohol Research and Education's (FARE) *2019 Annual Alcohol Poll: Attitudes and Behaviours* has provided a comprehensive snapshot of Australia's relationship with alcohol. This poll is reliant on people's opinions and perceptions of alcohol consumption in Australia.

The following table provides an overview of community attitudes indicating whether Australia believes that Australia has a problem with excess drinking or alcohol abuse between 2010 and 2019.

Shutterstock.com/Edge Creative

	2010 (%)	2011 (%)	2012 (%)	2013 (%)	2014 (%)	2015 (%)	2016 (%)	2017 (%)	2018 (%)	2019 (%)
Yes	73	80	76	75	78	75	78	78	73	66↓
No	16	14	15	14	12	15	12	12	16	22↑
Unsure	11	6	9	11	10	11	10	9	12	12

↑↓ denotes a significant change from the previous year's results (applied to 2019 data only).

1 Consider data in the table above. 66% of Australians surveyed in 2019 believe that Australia has a problem with excess drinking or alcohol abuse, while 22% do not believe this to be the case and 12% are unsure. This is down from 73% in 2018.

Discuss reasons why peoples' perceptions of alcohol abuse in Australia is changing for the better. Conduct an online search to help you formulate an answer.

____

____

____

____

____

____

2 Consider data in the table above. Explain the trend in peoples' perception of alcohol abuse since 2010.

____

____

____

____

____

____

From www.fare.org.au

Alcohol consumption in Australia is considered by many to be a socially acceptable 'drug'. Alcohol is the most widely used drug in Australia. While trends indicate that alcohol consumption habits are changing, around 1 in 5 (17.1%) of Australians over the age of 14 drink at levels that put them at risk of alcohol-related harm.

What do you already know about the effects of alcohol? Use your student book to annotate the diagrams showing the short- and long-term effects of alcohol.

Short-term effects of alcohol

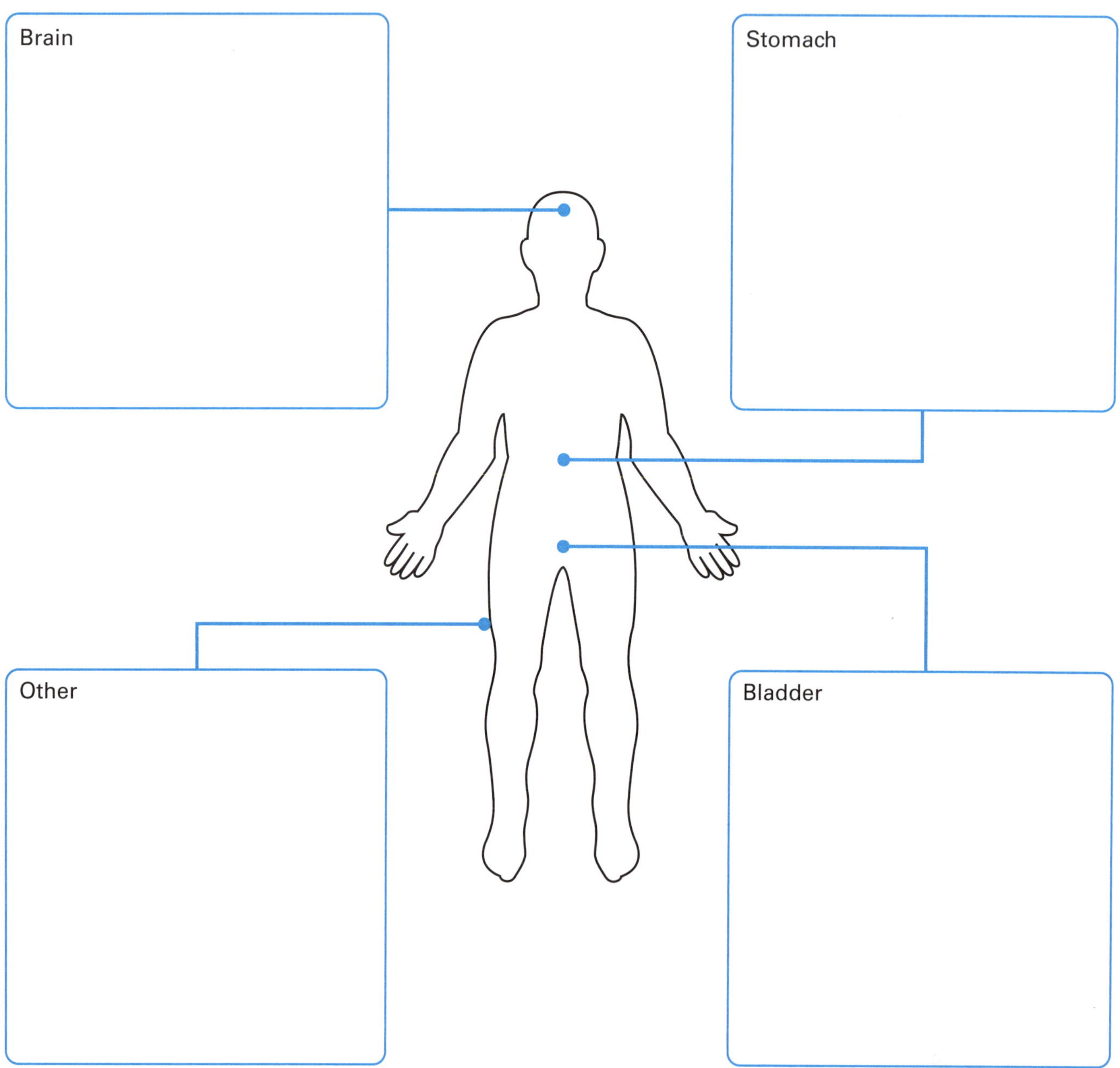

9780170465533

Long-term effects of alcohol

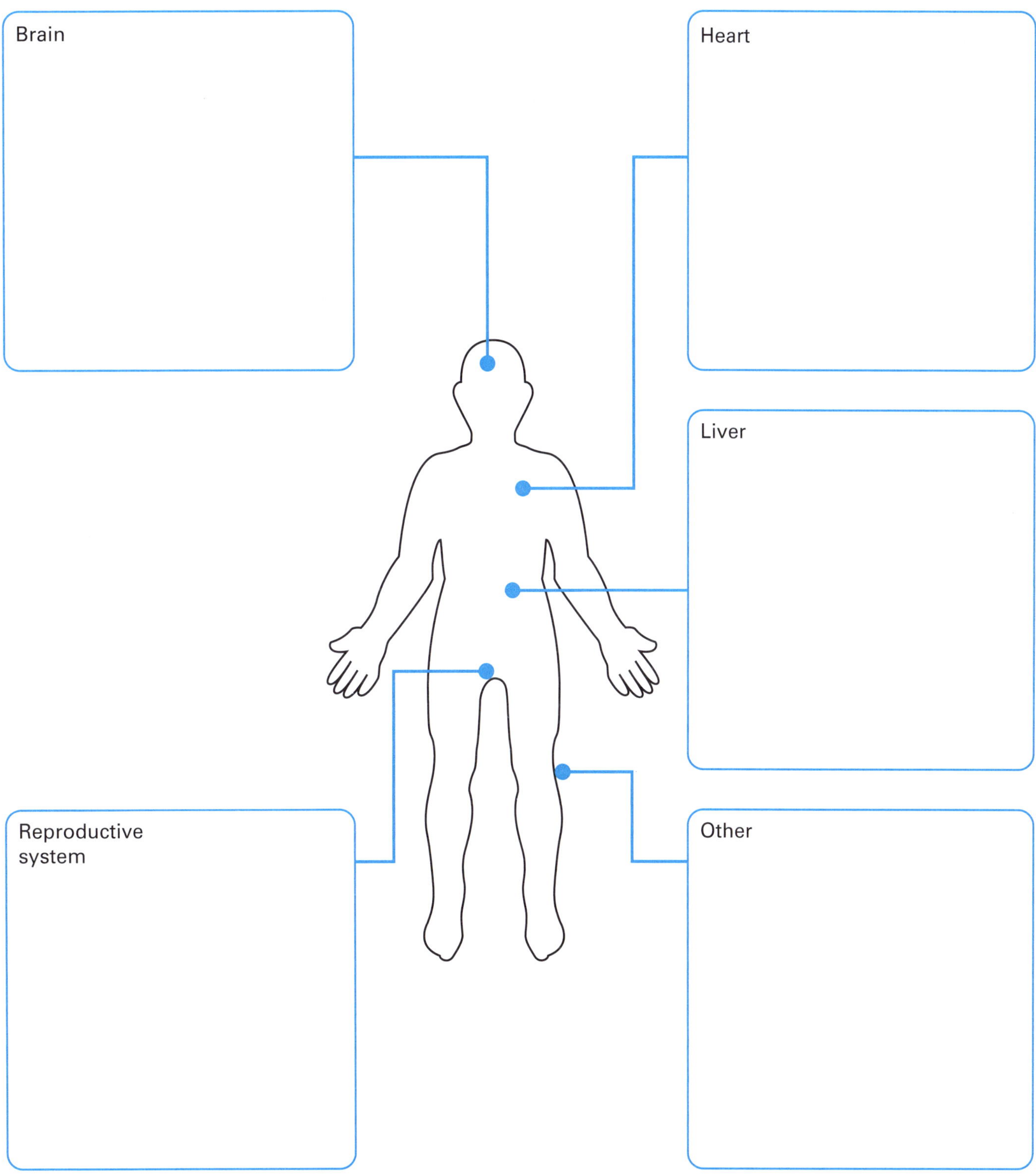

WORKSHEET 1.6 HARMS ASSOCIATED WITH ALCOHOL CONSUMPTION

Complete the table on the following page by listing possible 'harms' that may occur at each stage of drinking. Read the effects of alcohol at each stage and choose the harm that may best fit that particular stage of alcohol consumption. Choose the harms from the word bank below.

Word bank of possible harms

- Drink driving
- Having unprotected sex
- Losing consciousness
- Being sick
- Death
- Being injured
- Doing something you may regret
- Violence
- Upsetting someone
- Lying
- Possible trouble with police
- Being in trouble with parents
- Damaging property
- Out of control behaviour
- Poor decision making
- Getting in a car with a drunk driver
- Suffering from a hangover
- Vandalism
- Feeling ashamed
- Saying something you may regret
- Losing self-respect
- Coma
- Stupid behaviour
- Getting into a fight
- Being argumentative

Stage of alcohol consumption	Effects of alcohol (will vary among individuals)	Possible harms
Euphoria One or two drinks consumed	Poor concentration Happiness Relaxation Increase in confidence Poor judgement Problems with fine motor skills such as reading and writing	
Excitement A few more drinks consumed	Increase in confidence Sleepiness Poor reactions Uncoordinated movements Poor balance Blurred vision Fewer inhibitions	
Confusion And a few more consumed	Confusion Dizziness Staggering Blurred vision Slurred speech Heightened emotions (irrational/aggressive; affectionate/withdrawn)	
Stupor Even more drinks consumed	Difficulty moving Inability to stand Nausea Vomiting Need to sleep	
Coma **Death** More still consumed ...	Slowed heart rate Loss of consciousness Coma Death	

Adapted from 'Rethinking Drugs: You're in control – Student Workbook, Australian Government Department of Education and the Australian Brewers Foundation

WORKSHEET 1.7 ALCOHOL MYTHS AND FACTS

Pages 18–22

A myth is a false belief or idea that is often assumed by many to be true but is in fact not true. A fact is based on hard evidence and has been proved to be true.

Read the following statements about alcohol and decide whether you think each statement is a myth or a fact. Circle the correct answer and provide reasons to support your decision.

1 Alcohol is a stimulant. Myth / Fact

Reason for my answer:

____

____

____

____

2 All people react to alcohol in the same way. Myth / Fact

Reason for my answer:

____

____

____

____

3 If alcohol is legal, it can't be that harmful. Myth / Fact

Reason for my answer:

____

____

____

____

4 Coffee and a cold shower will help to make someone sober. Myth / Fact

Reason for my answer:

____

____

____

____

9780170465533

5 Drinking spirits will make you drunk faster than drinking beer or wine. Myth / Fact

Reason for my answer:

6 People drink alcohol to have a good time. Myth / Fact

Reason for my answer:

7 Drinking alcohol makes people put on weight. Myth / Fact

Reason for my answer:

8 Drinking alcohol is a great way to relax and unwind. Myth / Fact

Reason for my answer:

WORKSHEET 1.8 SMOKING: IT'S TIME TO BREAK THE HABIT!

Cost to society:

$137 billion!

Coronary heart disease is the leading cause of death in Australia and smoking is a major contributing risk factor. Approximately 20 000 Australians die each year from preventable smoking-related illnesses. The cost of smoking in our society is estimated to be more than $137 billion per year! Can you write that figure in numbers?

Pages 22–24

What do you already know about smoking? Use your student book to annotate the diagrams showing the short- and long-term effects of smoking.

Short-term effects of smoking

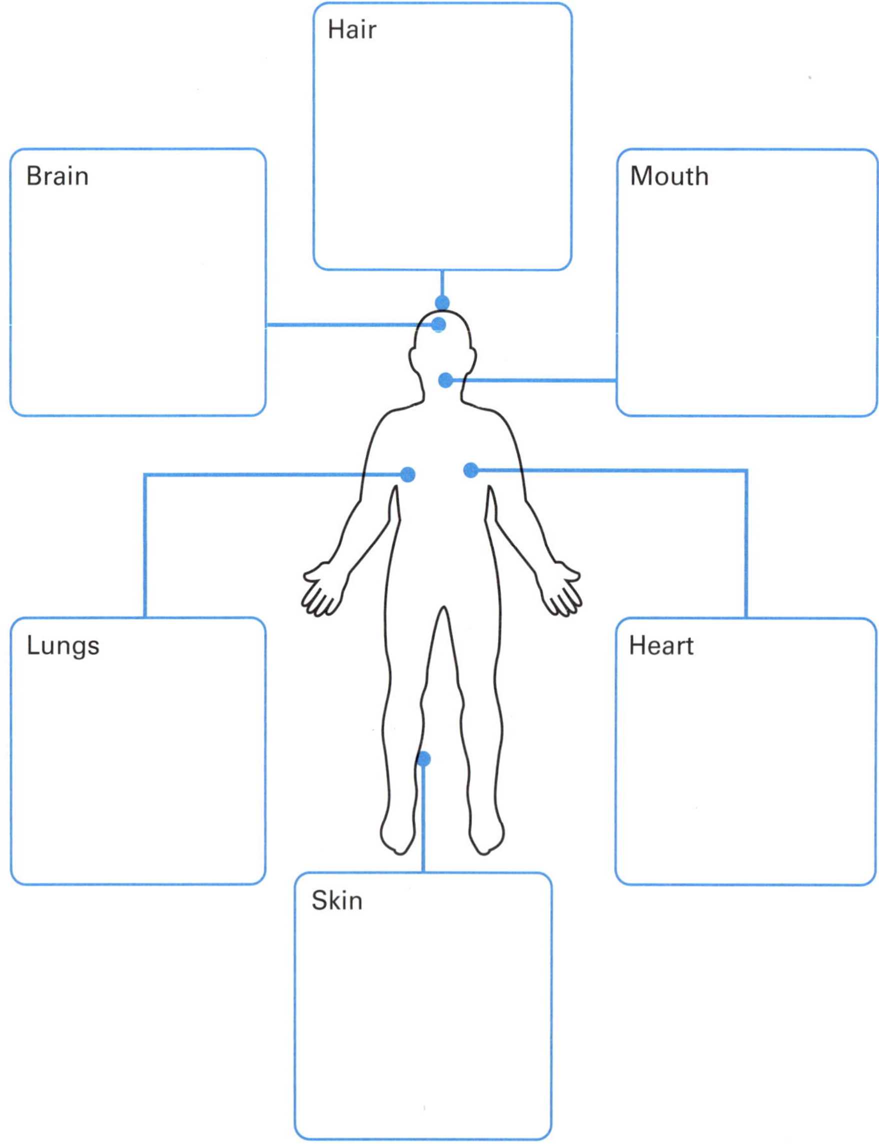

Long-term effects of smoking

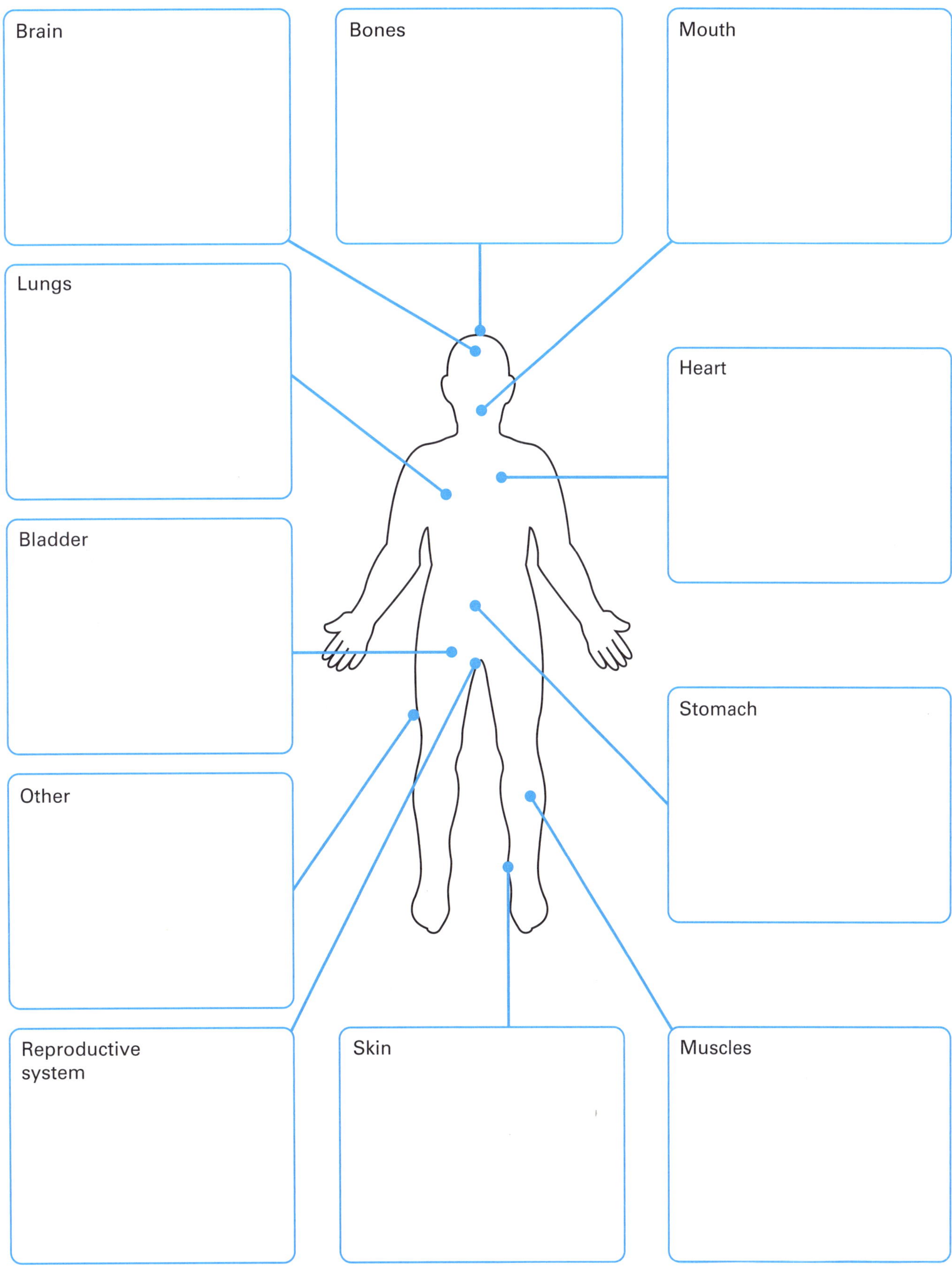

9780170465533

WORKSHEET 1.9 ELECTRONIC CIGARETTES

Pages 22–4

Electronic cigarettes, otherwise known as e-cigarettes or vapes, have emerged in the past 10 years as a popular alternative to smoking tobacco. E-cigarettes are hand-held, battery-operated devices that heat a liquid to produce a vapour that is then inhaled by the user. This method of simulated smoking is commonly referred to as vaping. Often marketed as an alternative method of quitting cigarette smoking, vaping also carries a set of risks and unknowns – especially for young people.

CASE STUDY

Marketed as a healthier alternative to cigarettes and, in its nicotine-free form, as harmless fun, vaping is a big deal for some ex-smokers, tobacco company executives, worried health regulators and, for better or worse, some kids. But, in recent weeks, alarm bells have been clanging in the United States over an outbreak of severe lung illnesses and the deaths of several people who vaped – amid an 'epidemic' of vaping in high schools.

Most vapes, or e-cigarettes, are electronic devices that use refillable tanks or disposable cartridges to house 'e-juice' – a concoction of chemicals and oils that can be combined with nicotine and flavourants. Instead of lighting the e-cigarette with a flame, vaping involves a battery being activated by inhaling or pushing a button that heats liquid-saturated wicks, often via metal coils, and turns the e-juice into an aerosol mist, which users inhale. Vapes can be filled with heated tobacco instead of e-juice.

The sale of liquid nicotine used in vaping is banned in Australia without a doctor's prescription. This is because it's classed as a 'dangerous poison'. The states have their own laws on the promotion, sale and use of vapes. For example, in NSW and Victoria vaping is banned in places where smoking is also banned but in Western Australia it is not. In Victoria and NSW, non-nicotine vapes can be sold to adults so long as the seller doesn't claim they help you to quit smoking. Western Australia and South Australia have banned the sale of products that are designed to look like cigarettes.

The concerns have led to the San Francisco-based company that controls about 70 per cent of the United States vaping market, Juul Labs, announcing it will no longer run TV, print or digital advertisements for its vaping devices and nor will it lobby against a proposed ban on vape flavours – the lolly-like fruit and dessert tastes that have helped spur the vaping craze. 'Juul was designed with adult smokers in mind,' it says, in caps, on its website.

In late September, South Korea and India – which has the second-most smokers in the world – became the latest countries to ban or warn about the sale of e-cigarettes. China, which has the world's most smokers, is having its own regulatory crackdown on vaping.

From Charlotte Grieve, 'What is vaping and is it bad for you?', *The Sydney Morning Herald*, 26 September 2019.

1 **Discuss** how e-cigarettes are similar to tobacco cigarettes and how they are different.

discuss to talk or write about a topic, taking into account different issues or ideas

AC

2 Juul Labs, a San Francisco-based company, currently controls 70 per cent of the US vaping market. Given the recent deaths and health concerns related to vaping, what has this company decided to do?

3 Consider what might have helped 'spur on' the vaping craze in the US.

4 Identify two countries that have the second highest number of smokers in the world and, given recent concerns about vaping, explain how these countries have tried to minimise the harm caused by vaping.

5 In your own words, define the term 'vaping'.

6 Consider reasons why the sale of liquid nicotine in vaping banned in Australia without a doctor's prescription.

7 Australian states have their own laws on the promotion, sale and use of vapes. **Recall** the two states which have banned the sale of products designed to look like cigarettes.

Recall present remembered ideas, facts, or experiences

AC

WORKSHEET 1.9 CONTINUED

8 Using the information in the case study, and other information you know, consider any facts that are **positive**, **negative** or **interesting** and write them in the table below.

Positive	Negative	Interesting

WORKSHEET 1.10 MAKE SMOKING HISTORY AND OTHER CAMPAIGNS

Weblink
Make smoking history

The 'Make Smoking History' campaign was established in 2000 with the goal to reduce smoking in Western Australia. Using the Make Smoking History website, answer the following questions.

1 What does the Make Smoking History campaign seek to do?

2 Click on the tab 'WHY QUIT' then 'PHYSICAL HEALTH' and answer the following questions:

a How many different smoking-related cancers are there?

b Identify and explain three smoking-related cancers.

c Identify and explain three other health effects caused by smoking.

3 Quitting smoking has significant health benefits. The moment you stop smoking, your body will begin to repair itself. Click on 'QUIT SUPPORT' then 'BENEFITS OF STAYING QUIT'. On the timeline below, identify the following health benefits a person would feel after their last cigarette.

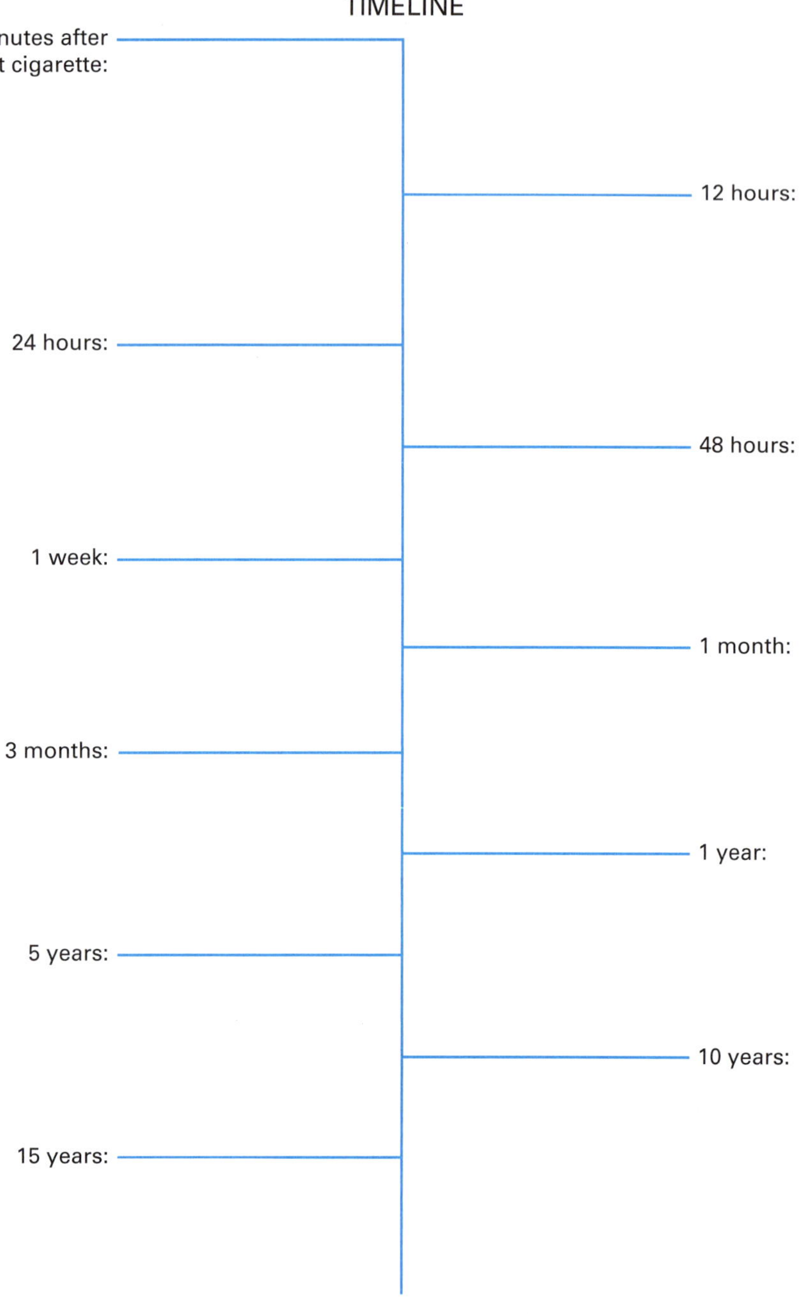

4 There are several mental health benefits of quitting smoking. Identify three mental health benefits.

____

____

____

5 Many people who smoke state that it 'relieves stress'. Discuss whether you agree with this justification.

____

____

____

____

____

PAST CAMPAIGNS

6 Click on 'QUIT STORIES', then '2018 TERRIE's TIPS' and read Terrie Hall's story. If Terrie was your friend in school, what 'ways to quit' advice would you have given her?

____

____

____

____

____

____

In the past, scare tactics have been used to deter people from smoking. Graphic and shocking images and statistics are often used to scare people to change a bad behaviour or habit.

AAP Images/AP

Click on 'SEE THE LATEST CAMPAIGN' on the Make Smoking History website and watch the commercial.

7 Is this latest commercial a scare tactic, designed to scare you to change a behaviour or are other tactics being used? **Explain** your answer.

explain
to provide information that demonstrates an understanding of a topic or something

8 Australia was the first country to sell cigarettes in plain packaging. Research this and assess whether you believe plain packaging is an effective way to raise people's awareness of the dangers of smoking.

9780170465533

WORKSHEET 1.11 SPELL CHECK

Your brain is amazing. It is like a powerful computer that controls the way you think and react. Did you know that the brain does not read every letter of a word by itself, but the word as a whole? This means that, even if the letters in a word are jumbled, you can still read the word correctly.

Read the following text and then translate it below. If in doubt, use your student book to help you. Time yourself and compare your time with a partner's.

Eegnry dnrkis

Smoe of the msot cmmoon sdie eefctfs acoisasetd wtih eegnry dnrik csupoitmnon icudlne:

- tmrreos
- piaaionttpls
- cshet pian
- dezsinzis
- isonnima.

WORKSHEET 1.12 JAMIE'S BMX COMPETITION

Read the case study below and answer the following questions.

CASE STUDY

Alamy Stock Photo/Louis-Paul st-onge Louis

Jamie had his sights set on an upcoming Freestyle BMX competition. He was in his element when he rode, always aiming for a higher level of achievement and living his dream. When Jamie was on his bike, he was constantly redefining his personal performance boundary, setting his limits and pushing them further. The thrill of extreme competition only heightened his adrenaline rush. The night before the competition, Jamie lay awake, thinking of his moves and couldn't wait to get on his beloved bike.

On his way to the competition he felt a little tired and didn't hesitate to buy a few cans of energy drink from the local store. The excitement of the competition had kept him from sleeping and he wanted to ensure he performed at his best. During the five hours of competition, Jamie consumed six cans of energy drink. He needed to be alert, focused and confident that he could pull off the stunts. Towards the end of the day, Jamie started to experience chest pain. He could feel his heart palpitating and he noticed he was sweating more than usual. The effects of the increased levels of caffeine consumption were starting to show. He did not realise that he was dangerously close to suffering a heart attack.

1 What was Jamie trying to achieve?

____

____

2 Identify the natural hormone secreted by the adrenal glands in times of major excitement.

____

propose to put forward (e.g. a point of view, idea, argument, suggestion) for consideration or action

summarise to give a brief statement of a general theme or major point/s

3 **Propose** reasons why Jamie chose to consume six energy drinks whilst competing.

____

____

4 **Summarise** the symptoms Jamie was experiencing and consider what could have gone wrong.

____

____

5 Decide on the advice you would give Jamie when he competes next time.

____

____

WORKSHEET 1.13 CANNABIS USE IN AUSTRALIA

Since 1984, students aged 12 to 17 years from across Australia have taken part in the Australian Secondary Students' Alcohol and Drug Survey (ASSAD). This survey is conducted every three years and students were asked about their lifetime and current use of alcohol and other drugs. Using the table below, answer the following questions on the lifetime use of cannabis among 12–17-year-old students from 1996 to 2017.

Prevalence (%) of lifetime cannabis use among 12–17-year-old students, 1996–2017

Survey year	12 years	13 years	14 years	15 years	16 years	17 years
1996	13	22	34	45	50	55
1999	9	17	28	38	44	50
2002	6	10	23	32	37	42
2005	5	9	15	23	30	32
2008	3	6	11	18	23	26
2011	3	6	11	17	25	29
2014	4	6	11	19	26	31
2017	3	6	10	22	29	34

Drug and Alcohol Research and Training Australia

Cannabis is the most used illegal drug in Australia. But even then, only 17% of 12–17-year-old students have used cannabis in their lifetime. 83% have never used cannabis!

1 Identify which year and age group saw the highest percentage of lifetime cannabis use among students.

____

2 Identify which year and age group saw the lowest percentage of lifetime cannabis use among students.

____

WORKSHEET 1.13 CONTINUED

3 Use the findings from the table to plot a line graph to show the lifetime cannabis use for each year group from 1996 to 2017.

describe to give an account of characteristics or features

AC

4 **Describe** the overall pattern of lifetime cannabis use among 12–17-year-olds since 1996. Provide possible reasons for this trend.

1

WORKSHEET 1.14 THINK IT THROUGH

Consider the effects of cannabis use. Read the following scenarios and answer the questions using facts and information from your student book or other research.

Page 33

SCENARIO 1

Two soccer players in the school team are nervous before competing in a major tournament. A friend offers the players a joint to calm their nerves.

1 Identify the risks involved for these two players.

2 Discuss how their actions could affect the whole team.

SCENARIO 2

A surfer is just about to hit the waves with his new board. His friend offers him a joint before he paddles out.

3 **Determine** the risks in this situation.

determine
to establish, conclude or ascertain after consideration, observation, investigation or calculation

AC

4 Consider whether the surfer should hit the waves. Research and use evidence to justify your decision.

WORKSHEET 1.15 EFFECTS OF CANNABIS

Page 33

Cannabis can have an adverse effect on people's lives, not only physically but mentally and socially. Conduct an online investigation on the effects of using cannabis. In your investigation, research the two areas below and write your results in the boxes provided.

Impact of cannabis on mental health	The social effects of cannabis

Investigation skills

WORKSHEET 1.16 MEDICINAL CANNABIS DEBATE: THE ADVANTAGES AND DISADVANTAGES

While medicinal cannabis is now legal in some parts of Australia, there is still much debate over whether medicinal cannabis should be prescribed by doctors to alleviate the symptoms of medical conditions.

SB Pages 33–5

In groups of four or five research online and prepare detailed notes on one of the below statements that will be allocated to your group by your teacher. Groups will then prepare a two-minute presentation to share their findings with the rest of the class and debate whether the statements are true or false. The statements to be allocated are:

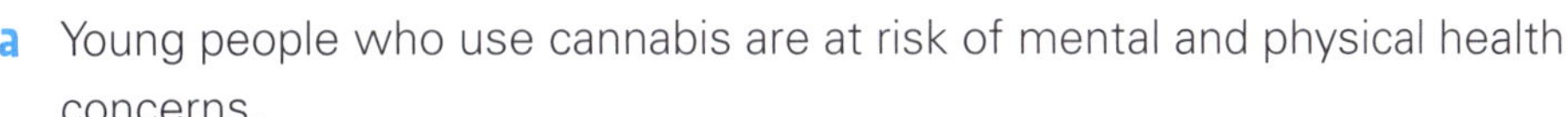

a Young people who use cannabis are at risk of mental and physical health concerns.

b Recreational cannabis is very different to medicinal cannabis.

c The effects of medicinal cannabis use are still unknown.

d Medicinal cannabis is legalised throughout Australia.

e Cannabis has been used for medicinal purposes for thousands of years.

f Medicinal cannabis is effective for pain management.

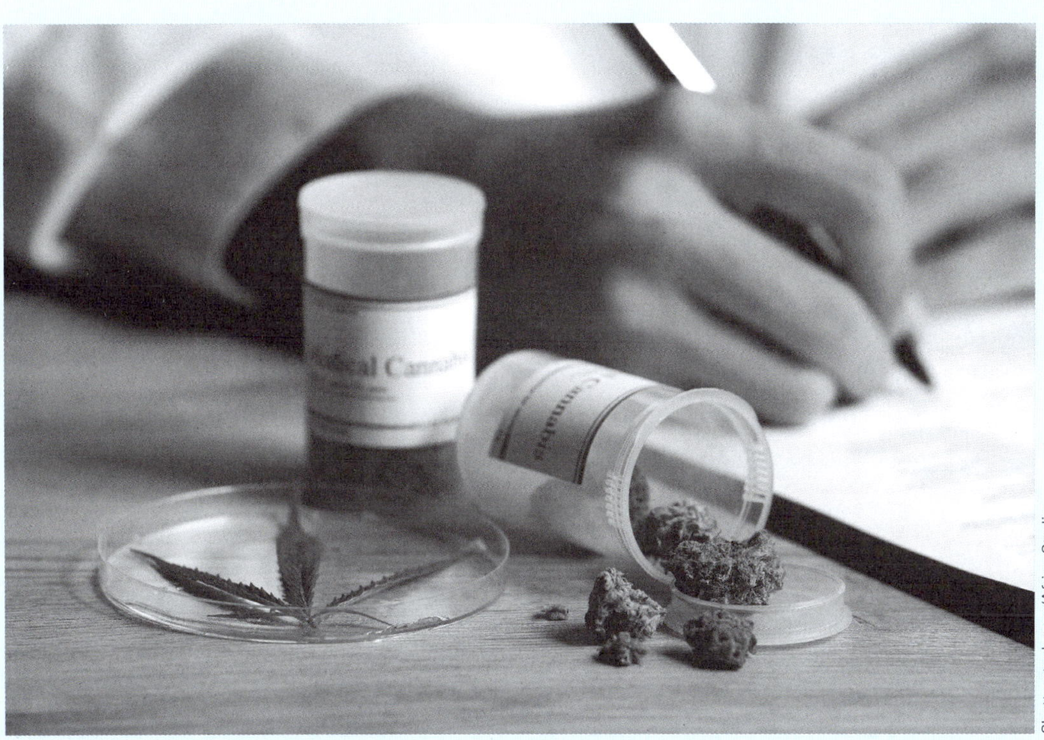

Shutterstock.com/Africa Studio

WORKSHEET 1.17 THE INCREASE IN ECSTASY USE

Read the news article below and answer the following questions.

CASE STUDY 1

The proportion of high school students using ecstasy has more than doubled in three years, prompting a leading drug educator to warn about the normalisation of the illicit drug's use among young people.

Paul Dillon, the founder of Drug and Alcohol Research and Training Australia, said the latest Australian Secondary School Students' Alcohol and Other Drug study revealed an 'alarming' increase in the consumption of ecstasy by students.

Mr Dillon said he was concerned about the blasé attitude of students towards MDMA (the main ingredient in ecstasy). 'All drugs have risks and the minute you don't have respect for drugs, you start doing things that are much more dangerous,' he said. 'We're going to see young people die.'

The survey of almost 20,000 high school students around Australia found 16 per cent of 17-year-old boys had tried ecstasy in 2017 compared to 9.2 per cent three years earlier.

The proportion of 17-year-old girls who had consumed the party drug increased from 4.7 per cent in 2014 to 9 per cent in 2017. Overall, the study found ecstasy use among students aged between 12 and 17 had increased from 2 per cent in 2011 to 5 per cent in 2017. It also found much higher rates of substance use by high school students with a mental health diagnosis.

Mr Dillon also expressed concern about the use of DIY pill-testing kits such as the EZ test to find out the contents of a pill, questioning whether young people were able to accurately interpret the results. He said the increased use of ecstasy by young people would lead to fatalities beyond the nightclubs and music festivals where drug reform campaigners and some politicians have called for the introduction of pill testing. Mr Dillon supports pill testing because it provides information about the contents of the drug that is tested, but he does not believe it is a 'silver bullet' to prevent festival deaths.

Mr Dillon said the study revealed three concerning drug trends among students: the normalisation of ecstasy, and increasing use of cannabis and inhalants such as nitrous oxide.

Melinda Lucas, a spokeswoman for the Alcohol and Drug Foundation, said the use of ecstasy by secondary students is low, but 'it is important for everyone to understand that is no safe level of drug use and any use increases the risk of harms such as accidents, injuries and overdose'.

'It is important young people understand that taking drugs is not the norm – only a small amount of high school students consume illicit drugs,' she said.

1 Describe the trends in ecstasy use across 12–17 year olds.

2 The 2017 Australian Secondary Students' Alcohol and Drug Survey found higher rates of substance use by which group of students?

3 Mr Dillon, while he supports the idea of pill testing as it provides information about the contents of the drug being tested, does not support DIY pill testing kits. Discuss reasons why.

4 Recall the three concerning drug trends revealed amongst students identified in the 2017 Australian Secondary School Students' Alcohol and Drug Survey.

5 What important message does Melinda Lucas, spokeswoman for the Alcohol and Drug Foundation, say about ecstasy use?

6 'Of concern is a 'blasé attitude of students towards MDMA'. Students seem to assume that MDMA is safe to use and therefore they can use more of it. Do you agree or disagree with this assumption? Investigate online to help you formulate your answer.

9780170465533

WORKSHEET 1.18 ROLE-PLAYS: WHAT WOULD YOU SAY?

Many teenagers like you may encounter one or more of the following scenarios at some point in their lives. Would you know how to respond if situations like these were to arise? In pairs, read over the five scenarios below and prepare two different responses for each scenario. Use facts to support your answer.

Once you are ready, practise your response by acting the scenario out to each other.

SCENARIO 1

You decide to go to your cousin's 15th birthday party. There are adults present. You meet your friend at the party and discover he has snuck in some beers for you to share. What will you say?

Response 1

____

____

____

____

Response 2

____

____

____

____

SCENARIO 2

A friend is around to watch a movie. She heads to the bathroom and finds the family medicine cabinet. You see her stealing the cough medicine and when you confront her, she says that 'no one will notice'. What will you say?

Response 1

____

____

____

____

Response 2

SCENARIO 3

You have never used an e-cigarette but your best friend has. He has had a bad chest infection in recent weeks, has a constant cough and is feeling very unwell. He continues to vape saying that vaping is not as dangerous as smoking. What will you say?

Response 1

Response 2

SCENARIO 4

Your friend is into motor cross and invites you to watch her compete at her local raceway. It's a hot day and the competition is fierce. She says she needs to keep hydrated and her 'energy levels' up. You notice she bought a pack of four energy drinks and she is drinking the final one. What will you say?

Response 1

WORKSHEET 1.18 CONTINUED

Response 2

SCENARIO 5

Your friend calls by to ask if you want to hang out that afternoon. You both head to the park to sit on the swings and chat. He then pulls out a small bottle of whisky from his jacket and he says, 'Try this. I got it from home. Dad won't notice it's gone because there are some other small bottles in the cabinet'. What will you say?

Response 1

Response 2

9780170465533

WORKSHEET 1.19 FACTORS INFLUENCING DRUG USE

Page 39

Young people choose to take drugs for a variety of reasons. They are influenced by three main factors: personal, environmental and social factors. Understanding these factors and the risks associated with drug use will help you to make responsible, safe and informed decisions.

1 Use the blank middle column to match the correct examples with one of the three main factors influencing drug use.

Personal		Peer pressure, role models and socio-economic background
Social		Family members, family conflict and education
Environmental		Stress and self-esteem

2 Identify five reasons why young people may use drugs.

3 Identify five reasons why young people do not use drugs.

4 Consider three healthier alternatives young people may participate in, instead of taking drugs.

WORKSHEET 1.20 HELP SEEKING STRATEGIES

Seeking help is key to ensuring the safety of yourself and others. Knowing when to seek help, how to access help, the type of assistance you may need and how to problem solve specific situations you may find yourself in and skills that will help you make effective choices and decisions. Answer the following questions relating to seeking help. Think back to when you have needed to ask for help.

1 When was it ok to ask for help?

____

____

2 Were there occasions when you found it challenging to ask for help?

____

____

3 How did you work out who to seek help from?

____

____

4 If you had a drug or alcohol related issue, where do you think you would go to seek help?

____

____

Seeking help is a process and there are a number of steps to consider prior to seeking assistance:

Step 1 Recognising there is a problem and that help is needed

Step 2 Consider the best person or place to get help

Step 3 Approach the person

Step 4 Explain how you feel

Step 5 Identify the problem

Step 6 Requesting help

Source: Adapted from https://www.sdera.wa.edu.au/media/3925/sdera-drug-talk-teacher-resource-help-seeking.pdf

Read the following scenario and outline how you would seek help using the help seeking process.

SCENARIO 1

I'm 14 now and I started taking Mum's codeine tablets about a year ago. I'm a mad rugby league fanatic, and I am always outside practising my passing and kicks with my mates but I remember the day I twisted my ankle well. It had been raining and the oval ball was wet. I wore sports shoes that had no grip and slid into my friend who landed heavily on my ankle. The pain was awful and the swelling insane. I had a game that weekend. I knew Mum had recently been to the doctors and was taking some stronger painkillers for a back complaint. They really helped. Whenever I injure myself now I always dip into her supplies. I think she knows as she mentioned the other day that she's noticed her tablets disappearing. I said she was just getting forgetful. I'm finding it hard to stop and am not sure what I should do. They make me push harder and mask the pain.

____

____

____

____

SCENARIO 2

My brother is older than me, I'm 15. At 18, he's legally allowed to drink alcohol. Whenever his friends are around, I always ask him if I can grab a bottle of beer. He never says no. When he's not at home, I've found myself going into his room and taking some of his beer without him knowing. It's becoming a bit of a regular habit and I'm sure I'm going to get caught. I drink by myself in my room. I'd rather drink than socialise with my friends. Sometimes I've found it difficult to wake in the morning, I'm often late for school and just not focusing in class. I know what I'm doing is wrong but am not sure what to do.

WORKSHEET 1.21 KNOCK-OUT QUIZ

In each corner of your classroom you will find letters a, b, c and d. This is a knock-out quiz. Once each question is read out, you are to move to the correct corner. Students who answer correctly will remain standing. Those who answer the question incorrectly are to return to their seats.

1 Which of the following is not a drug?

a caffeine b alcohol c ecstasy d grape juice

2 Which drug is the most socially accepted in Australian society?

a tobacco b alcohol c cannabis d ecstasy

3 What is the legal age in Australia to buy cigarettes?

a 16 b 18 c 21 d 25

4 Which drug is the 'odd' one out?

a LSD b alcohol c cannabis d opiates

5 Which drug is the most widely used legal psychoactive drug?

a alcohol b tobacco c caffeine d cannabis

6 Which telephone number would you dial if you required emergency assistance?

a 911 b 999 c 000 d 991

7 Which of the following is commonly known as ecstasy?

a MDA b MDD c MDAD d MDMA

8 The number of chemicals found in a cigarette is around:

a 70 b 700 c 7000 d 70 000

9 Which symptom is not typical of a hangover?

a headache b hearing loss c anxiety d muscle aches

10 Which substance is responsible for more deaths due to toxic overdose than any other substance?

a tobacco b alcohol c caffeine d energy drinks

WORKSHEET 1.22 WHY SOME ATHLETES USE PERFORMANCE-ENHANCING DRUGS

Using your student book or other research, identify the reasons why athletes may use performance-enhancing drugs and label them onto the diagram below.

SB
Pages 51–4

1 In your own words, **define** the term 'doping'.

____

____

define to give the meaning of a word, phrase, concept or physical quantity

2 Can you think of ways in which athletes have enhanced their sporting performance by doping?

____

____

____

____

____

3 Can you think of any athletes who have been caught doping?

____

____

____

____

Investigation skills

WORKSHEET 1.23 PERFORMANCE-ENHANCING DRUGS INVESTIGATION

Athletes can enhance their performance to gain a winning edge by using *legal* methods. In groups of four, choose one legal substance from your student book to investigate further. Answer the following four focus questions based on the legal substance of your choice.

1 In which sports is the substance used, and why?

____

____

____

____

____

2 Explain how this substance enhances sporting performance.

____

____

____

____

____

3 Discuss the health risks associated with using this substance.

____

____

____

____

____

4 Investigate alternative healthier choices that can enhance performance in the same way as your chosen substance.

____

____

____

____

____

9780170465533

WORKSHEET 1.24 GLOSSARY TERMS

Some key terms about drugs and alcohol are listed in the first column of the table below. Complete the rest of the table as follows:

- Fill out the second column with your own definition of the word.
- Complete the third column with the definition used in the student book or a dictionary.
- Complete the fourth column with a sentence that uses the word in the context of this chapter.

Glossary term	What I think it means	What the student book or dictionary says it means	How I can use it in a sentence about alcohol and other drugs
Drugs			
Psychoactive			
Drug abuse			
Narcotic			
Medicines			
Carcinogen			
Overdose			
Electrolyte			
Holistic			
Ergogenic aids			
Synthetic			

CHAPTER 1 REVIEW

It is now time for you to have your say. Write a response to the following questions about alcohol and other drugs.

1 State the decision your parents or caregivers would like you to make about drinking alcohol and using other drugs.

2 Consider how friends and peers might influence your decisions when it comes to using alcohol and other drugs.

3 What would you do if you were offered a cigarette?

4 Do you think your decision will change as you get older? Discuss reasons for your answer.

9780170465533

2

EAT WELL, LIVE WELL

WORKSHEET 2.1 BRAINSTORM: WHAT DO YOU KNOW ALREADY?

Page 60

Let's find out how much you already know about food, nutrition and how to eat healthily. Use this working space to reflect upon and brainstorm all the relevant terms you know about this topic. Make sure you have a title in the middle of your brainstorm. You may be able to use different colours or highlighters to 'cluster' similar terms together. Share your brainstorm with the class, and make sure you add any suggestions from other students that you may have missed.

9780170465533

WORKSHEET 2.2 FOOD GROUPS

1 Match the standard food servings listed below with their correct food group in the table. Most other foods you might choose to eat can be sorted in this way.

SB Page 61

A cup is 250 mL, which is less than a typical can of soft drink (375 mL).

½ cup corn
1 apple
½ potato
2 apricots
½ cup pasta
3 crispbreads
170 g tofu

1 slice bread
2 eggs
1 cup cooked kidney beans
½ cup rice
1 cup canned fruit
1 cup raw salad vegetables
2 slices vegan cheese

¾ cup yoghurt
1 cup milk
2 slices cheese
½ cup cooked green vegetables
80 g cooked chicken
65 g cooked lean meat
1 cup soy milk

Vegetables	Fruit	Grains (cereals)	Meat, fish, eggs, nuts, legumes, tofu	Milk, yoghurt, cheese, non-dairy alternatives

Investigation skills

WORKSHEET 2.3 WHO MEETS THE GUIDELINES FOR SERVES PER DAY?

Conduct a survey of your class to determine how many people are meeting the recommended number of serves per day of each of the food groups. The recommended number of daily serves for 12 to 13 year olds are:

- vegetables and legumes/beans: 5
- fruit: 2
- grains and cereals: 5–6
- meat, fish, eggs, nuts and tofu: 2
- milk, yoghurt, cheese and non-dairy alternatives: 3.

Pages 61-4

1 Using the table below, find how many of your classmates met these recommendations yesterday.

Servings	Number in class who met this
5 servings of vegetables, legumes/beans	
2 servings of fruit	
5–6 servings of grains and cereals	
2 servings of meat, fish, eggs, nuts	
3 servings of milk, yoghurt, cheese	

2 Complete the graph below to illustrate your results.

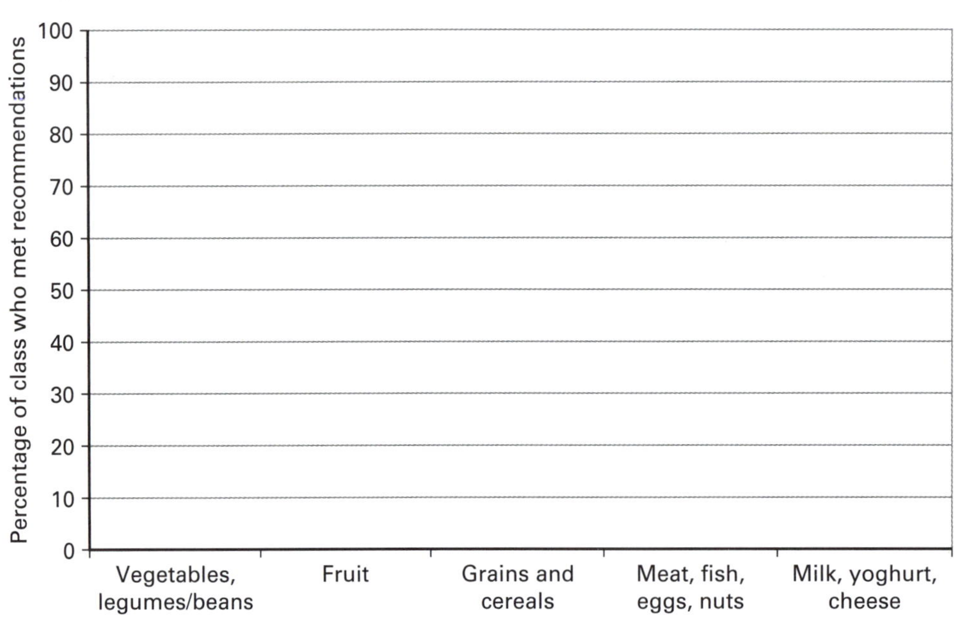

9780170465533

WORKSHEET 2.3 CONTINUED

3 **Evaluate** how well your class is meeting the recommended amount of serves per day of each food group. Describe any obvious issues.

evaluate to examine and judge the merit, significance or value of something

4 Which food group was most likely to have its recommended number of serves met? Provide reasons why you think this is so.

5 **Discuss** how likely it is that the results of your survey would match that of a survey of the wider community or population.

discuss to talk or write about a topic, taking into account different issues or ideas

WORKSHEET 2.4 SERVES PER DAY – FILL IN THE GAPS

Page 67

This table shows the recommended average daily number of serves of each of the food groups. There are a few missing from the table. Using the rest of the information in the table, you may be able to distinguish some patterns that will enable you to fill in the missing numbers.

Recommended average daily serves from each of the five food groups

	Vegetables and legumes/beans	Fruit	Grain or cereal foods	Lean meat, poultry, fish, eggs, nuts, tofu and seeds	Milk, yoghurt, cheese and non-dairy alternatives	Approx. number of additional serves from the five food groups or discretionary choices
Toddlers						
1–2	2–3	½	4	1	1–1½	
Boys						
2–3	2½	1	4	1	1½	0–1
4–8	4½	1½	4	1½		0–2½
9–11		2	5	2½	2½	0–3
12–13	5½				3½	0–3
14–18	5½	2	7	2½	3½	0–5
Girls						
2–3	2½	1	4	1	1½	0–1
4–8	4½	1½		1½	1½	0–1
9–11	5	2	4	2½	3	0–3
12–13			5			0–2½
14–18	5	2	7	2½	3½	0–2½

Based on material provided by the National Health and Medical Research Council

WORKSHEET 2.5 MAKING A DECISION

Use this decision-making model to identify and address an issue you face with your diet or the factors that limit your healthy eating options.

Page 70

Issue
Identify a problem or barrier you face to having a healthy diet.

Choices

Positive consequences

Negative consequences

Decisions
I will ...

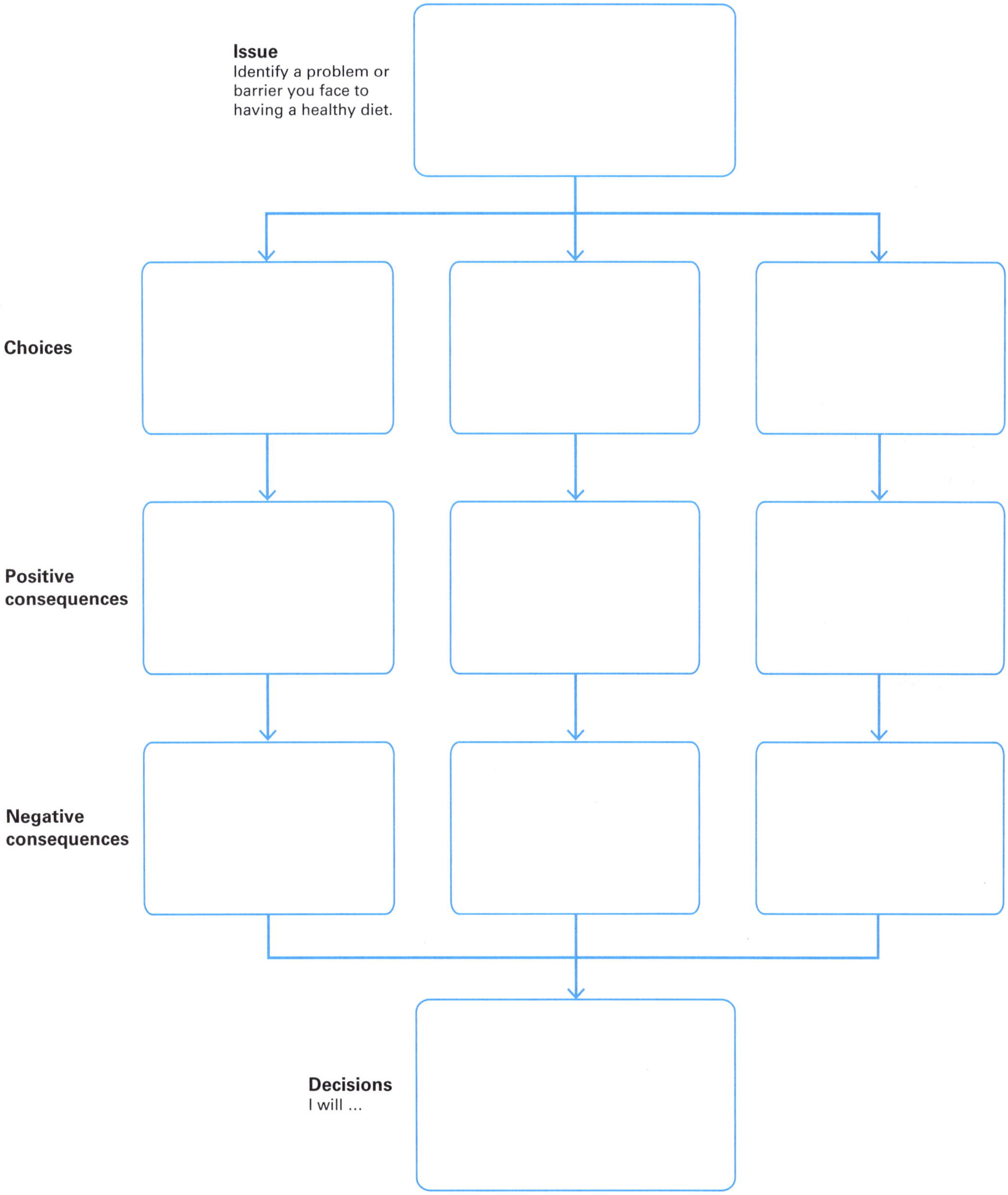

WORKSHEET 2.6 ESTIMATING YOUR ENERGY NEEDS

Page 79

You can make a close estimate of your own daily energy needs. You would need to monitor this over time to see if it is accurate, and bear in mind the need to make adjustments if your activity levels change, for example, due to injury.

There are several versions available online, but be careful about using reliable sources and up-to-date information. It is best to find a site that uses kilojoules, such as the ones from the Australian Government's National Health and Medical Research Council and the Department of Health. Search for the Eat for Health energy calculator online.

This calculator requires your age, gender, weight and an estimate of your physical activity level (PAL). You can work out your PAL from the guide provided: it will be between 1.2 and 2.2. Most people who are reasonably active are 1.6 to 1.8.

- **1.2:** bed-ridden
- **1.4:** very sedentary, no physical activity
- **1.6:** light activity, such as walking, no strenuous activity
- **1.8:** moderate activity, little sitting down
- **2.0:** heavy activity, such as manual labour or strenuous activity
- **2.2:** vigorous activity, such as elite athletes

Enter your own information and record the estimated energy intake you should be aiming for. Now experiment by changing the variables (e.g. age) and seeing how the answers differ. Make sure you only change one variable at a time and change it back to your original input before changing something else.

Your estimated energy requirements: ______________ kJ

Changing the ______________ to ______________ changes the energy estimate to: ______________

Changing the ______________ to ______________ changes the energy estimate to: ______________

Changing the ______________ to ______________ changes the energy estimate to: ______________

reflect on to think about deeply and carefully

AC

Try inputting the age, gender, weight and activity levels of different family members to see how these variables impact their approximate energy needs. **Reflect on** their eating habits and establish whether or not these different energy needs are reflected in their eating habits.

____

____

____

____

____

9780170465533

WORKSHEET 2.7 LETTERS TO AN EXPERT

Using your expertise on nutrition, respond to these queries to a 'Nutrition Expert' (that's you!) below by providing advice to achieve a healthy outcome.

Dear Nutrition Expert,

I'm 13 years old and have just started high school. I'm finding it to be very busy, with so much to do and so much moving around the school for different lessons and activities. I want to do my best and I try really hard, but I just feel so flat most of the time.

I snack regularly throughout the day, but I feel like I am hungry again half an hour after eating. Once lunchtime comes around I feel really tired, and my friends wonder why I don't want to play. I don't think I will bother trying out for the after-school sports teams. What can I do to keep my energy levels up for the whole day?

Ysobel

WORKSHEET 2.7 CONTINUED

Dear Nutrition Expert,

My friend has been preparing for her final exams this year. Study is taking up so much of her time that she has stopped playing sport and exercising. She has been eating a fair bit of fast food lately because of her late-night studying, and also before or after her shifts at her casual job. I have noticed she is putting on some weight and seems to be tired and sick all the time. When I ask her about it she usually brushes me off, but I know that she worries about it. I'm concerned some people might start giving her a hard time, and I know her coach isn't going to be happy when she heads back to training. Is there a way I can help her?

Regan

9780170465533

Dear Nutrition Expert,

I'm a 15-year-old guy and no matter what I do, I can't seem to bulk up. All the other guys my age are way bigger than me, and even though I am pretty fit and active, my lack of size (muscle!) is starting to affect my performance in games and competitions. I see some guys at training and coming out of the local weights gym drinking protein shakes and taking supplements, but these seem really expensive. My best mate reckons that he started getting bigger by eating lots of high energy food (especially junk food). I think I might try this strategy too. Will this work?

Ahmad

WORKSHEET 2.8 TEMPTATION

Pages 82–4

Sometimes we have the best intentions to make healthy eating choices but find it too hard to say 'no thanks' when we are tempted to make less-healthy choices. As with any skill, rehearsing it beforehand can make it easier to do in the future. Practise now in pairs or small groups.

create to produce or evolve from one's own thought or imagination

devise to think out; plan; contrive; invent

1 **Create** a scenario in which a person or a few people are offered or tempted by some unhealthy food choices.

2 **Devise** a script that conveys how the offer or temptation of the unhealthy food was declined.

3 Allocate roles and rehearse as a pair or group, and be ready to perform to your class when asked by your teacher.

For each of the statements below, write its number on the continuum at the point that represents how much you agree or disagree with the statement.

1 Fast food/junk food advertising should not be permitted during children's or family TV programs.

2 Healthy eating is not just about being overweight or a healthy weight, but also about giving your body the nutrients it needs.

3 Fast food/junk food companies should not be allowed to sponsor sporting events.

4 A 'fat tax' on unhealthy foods to make them more expensive would be an effective strategy to reduce obesity.

5 Eating healthily seems to be difficult, time-consuming and/or expensive.

6 We can eat both healthily and sustainably.

Strongly disagree	Disagree	Neither agree nor disagree	Agree	Strongly agree

9780170465533

WORKSHEET 2.9 YOUR OPINION

Justify your positioning of each of the statements below.

justify
to show how an argument or conclusion is right or reasonable

1

2

3

4

5

6

WORKSHEET 2.10 USING KEY TERMS

Construct a sentence that includes each term listed below, using it in an appropriate context. Try to put a positive angle on your sentences.

Extension: Instead of using separate sentences for each term, try to develop a single passage of text that allows you to use all the terms. It might help if you picture yourself giving a presentation or speech at a school assembly, environmental convention or community event.

- Ecological footprint
- Organic food
- Landfill
- Sustainability
- Bush tucker
- Tap water

construct to create or put together (e.g. an argument) by arranging ideas or items

WORKSHEET 2.11 CHASE THE RAINBOW

List as many fruits and vegetables as possible under each colour heading.

Option 1 – compare with your class mates, who can make the longest list using just their own knowledge, no outside help or internet research.

Option 2 – compare with your classmates, who can make the longest list using other resources such as internet research.

Yellow/ Orange	Green	Red	Blue/Purple	White/Brown

How many of these fruits and vegetables have you eaten in the last week?

WORKSHEET 2.12 COMPONENTS OF HEALTH

There are many benefits to having a healthy diet. Using the graphic organiser below, describe some ways that a healthy diet can benefit us in each 'component of health'.

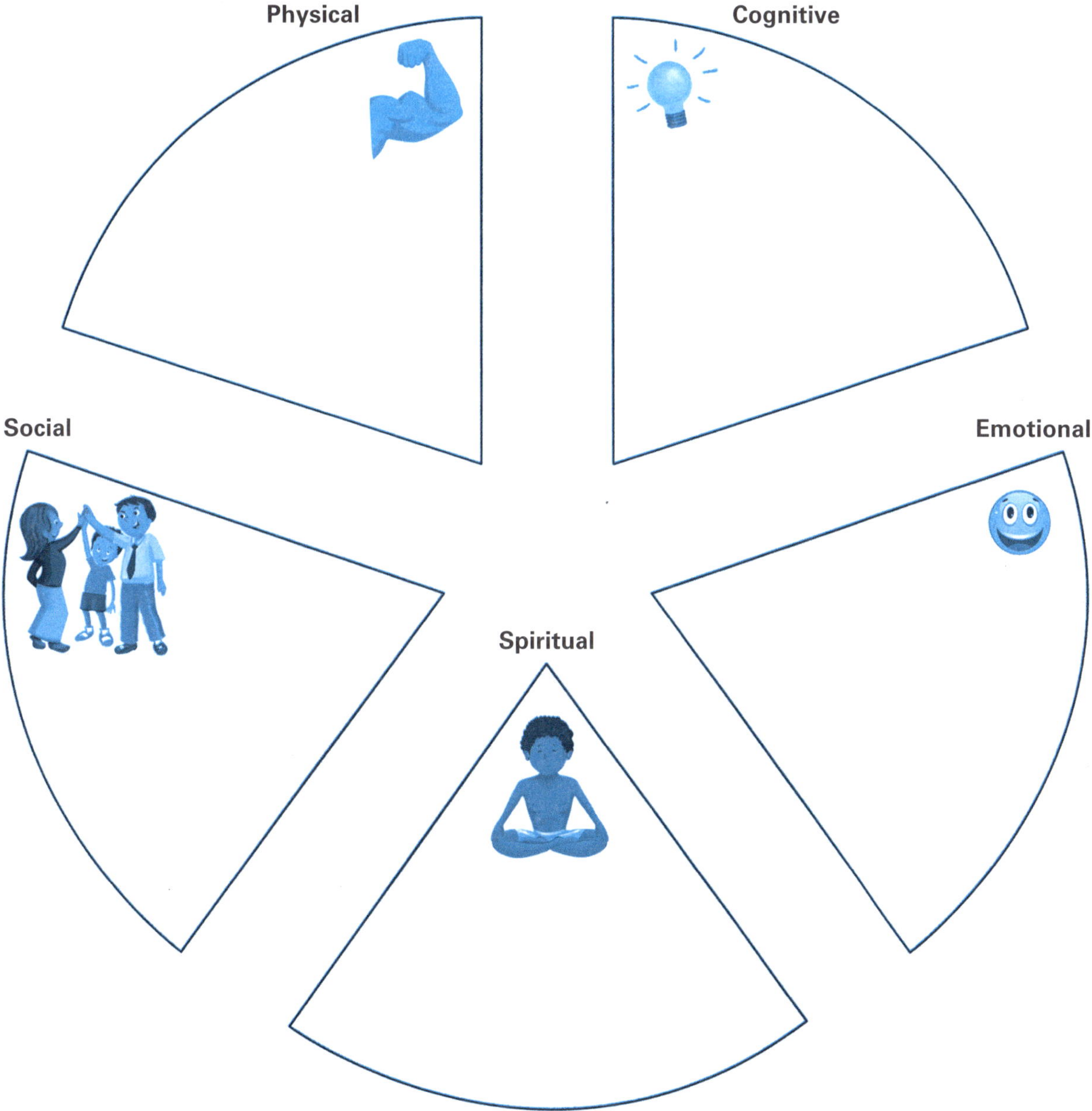

CHAPTER 2 REVIEW

Reflect upon your learning in this chapter. Completing the following sentences will help to structure your thoughts.

1 I developed a particular strength in:

2 I was surprised to discover:

3 Information I found useful:

4 I am looking forward to:

5 I hope more people get the chance to:

6 I can see my lifestyle changing:

9780170465533

3

HEALTH BENEFITS OF PHYSICAL ACTIVITY

WORKSHEET 3.1 PHYSICAL ACTIVITY IS …

For each of the statements below, write its number on the continuum at the point that represents how much you agree or disagree with the statement.

There are no right or wrong answers.

1 Every child (5–18 years) should play at least one competitive sport, either at school or in their local community.

2 Physical activity is important throughout all stages of a person's life.

3 Physical activity is important for all aspects of health: physical, social, emotional, cognitive and spiritual.

4 Physical education should be compulsory in all years of schooling and not stop at the end of Year 9 or 10.

5 Physical activity is only important if you are overweight or unhealthy.

Strongly disagree	Disagree	Neither agree nor disagree	Agree	Strongly agree

justify to show how an argument or conclusion is right or reasonable

AC

Justify where you placed each of the statements.

1 ______________________________

____

____

____

____

____

____

2 ______________________________

____

____

____

____

____

____

9780170465533

3 ____

4 ____

5 ____

WORKSHEET 3.2 ACTIVE TRANSPORT

Investigation skills

SB Page 108

Active transport is a domain of physical activity and refers to physical activity that results in reaching a destination.

ACTIVE TRANSPORT: CRITICAL POLICY BRIEF

The benefits of shifting from private car travel to active transport modes, such as walking and cycling, are well recognised and have been promoted by both national and state governments. Yet, active transport rates in Australia remain low in comparison with many European and Asian countries. Especially concerning is the decline in active transport among children, with less than a third of Australian children now regularly walking or cycling to school.

Getty Images/Education Images/ Universal Images

Overview

The benefits of active transport are recognised in health, transport and urban planning fields. Active transport reduces congestion in the road network and can reduce infrastructure costs, as well as delivering health benefits through physical activity and disease prevention. In Brisbane, for example, active transport has been estimated to gain 33,000 health-adjusted life years by 2026, generating net savings of $183 million.

'Globally, physical inactivity causes 3 million deaths per year. One of the most effective means of increasing physical activity is through urban planning and transport policies'

(World Health Organization, 2009).

Active transport delivers environmental benefits by contributing to lower air and noise pollution, with positive flow-on health outcomes. Infrastructure that supports walking and cycling also contributes to social equity and inclusion goals, providing opportunities for low-cost modes of travel.

However, active transport uptake in Australia remains very low in comparison with many European and Asian countries, with only 4% of the Australian workforce commuting by walking and 1% by cycling.

The number of Australian children walking or cycling to school has halved over the past 40 years, with less than a third now regularly walking or cycling to school.

A comprehensive cross-sectoral strategy is needed to increase the number of Victorians using active transport modes. This policy brief highlights the need to: build walkable neighbourhoods; further develop proximity-based planning policies; increase investment in cycling infrastructure and education; and coordinate active transport and public transport provision.

Build neighbourhoods that encourage active transport

A walkable neighbourhood encourages local living, with people being able to safely and conveniently walk or cycle to their preferred destinations. Walkable neighbourhoods have high residential density and a well-connected, safe pedestrian street network. Higher residential densities provide the foundations for well-serviced public transport infrastructure and locally accessible destinations, goods and services. However, dwelling densities in Melbourne remain low and Plan Melbourne 2017–2050 has set a relatively unambitious density target of 15 dwellings per hectare. To realise the goal of creating a city of walkable, 20-minute neighbourhoods, a residential density target of at least 25 dwellings per hectare is needed.

Urban planning and transport policies are strong and direct mechanisms to spur active transport. Proximity-related planning policies in Victoria, such as mandated distances to supermarkets and public transport stops, promote integrated planning that encourages uptake of active transport. Universal access to active transport is supported by infrastructure such as footpaths and cycle lanes around public transport stops, retail precincts and employment hubs. Car park pricing and availability policies can also discourage private vehicle use where public transport services are available. It is important that these policies guide the development of new residential housing and activity hubs, and that their implementation is monitored to assess outcomes.

Invest in Cycling Infrastructure and Education

Australia's low rates of commuter cycling reflect a substantial under-investment in cycling infrastructure. Promoting cycling as a convenient, healthy and safe travel mode requires development of connected bicycle networks and improved links to existing cycle paths. Cycling can be made more accessible for people of all ages and abilities by providing separated cycle lanes on major cycling corridors. Traffic calming features such as separated cycle lanes and controlled crossings enhance cyclist safety, which is especially important in encouraging cycling among younger and older riders. In Australia, Canberra has been most successful in increasing commuter cycling rates, achieving a 15% increase in cycling to work between 2011 and 2016.

This is supported by an integrated Active Travel Framework, an Active Travel Office to coordinate policy implementation, and major investment in cycling-related infrastructure and education.

Coordinate Active Transport Infrastructure and Public Transport Provision

Integrated local and regional planning enhances mobility and access to destinations, improving social equity and health outcomes. Active transport and public transport are complementary, with the majority of public transport journeys involving walking or cycling. The provision of high-quality public transport with active transport connections increases the distances accessible by modes other than car. There are opportunities to promote active transport aligned to public transport use by

WORKSHEET 3.2 CONTINUED

developing walking and cycling infrastructure around public transport facilities. This involves planning safe, connected cycle and walkways to public transport services, as well as developing urban green spaces along walking and cycling pathways.

From Dr Hannah Badland and Dr Claire Boulange, RMIT Centre of Urban Research: https://cur.org.au/cms/wp-content/uploads/2018/11/active-transport-policy-brief.pdf

1 Traffic calming features such as separated cycle lanes and controlled crossings enhance cyclist safety, which is especially important in encouraging cycling among younger and older riders. Investigate and summarise how many of these features appear in your capital city and local neighbourhood.

2 What could local councils, workplaces and schools do to encourage more people to use active transportation? List at least two possible initiatives for each.

Local councils:

Workplaces:

Schools:

9780170465533

WORKSHEET 3.3 HOW HARD ARE WE WORKING?

There are a number of ways to calculate how hard you are working during various physical activities.

One method is to measure your heart rate. Here are five methods you can use to measure your heart rate:

- taking your pulse (wrist or neck)
- using a heart rate monitor (can be on a smart-watch, or personal tracker)
- using an app on a smart phone or tablet
- having an electrocardiogram (ECG)
- RPE (rate of perceived exertion).

Compare and contrast any three of these five methods using a Venn diagram. Remember to say what is the same about each method (where the circles overlap) and what is different (where there is no overlap). The shaded area in the middle has been filled in for you.

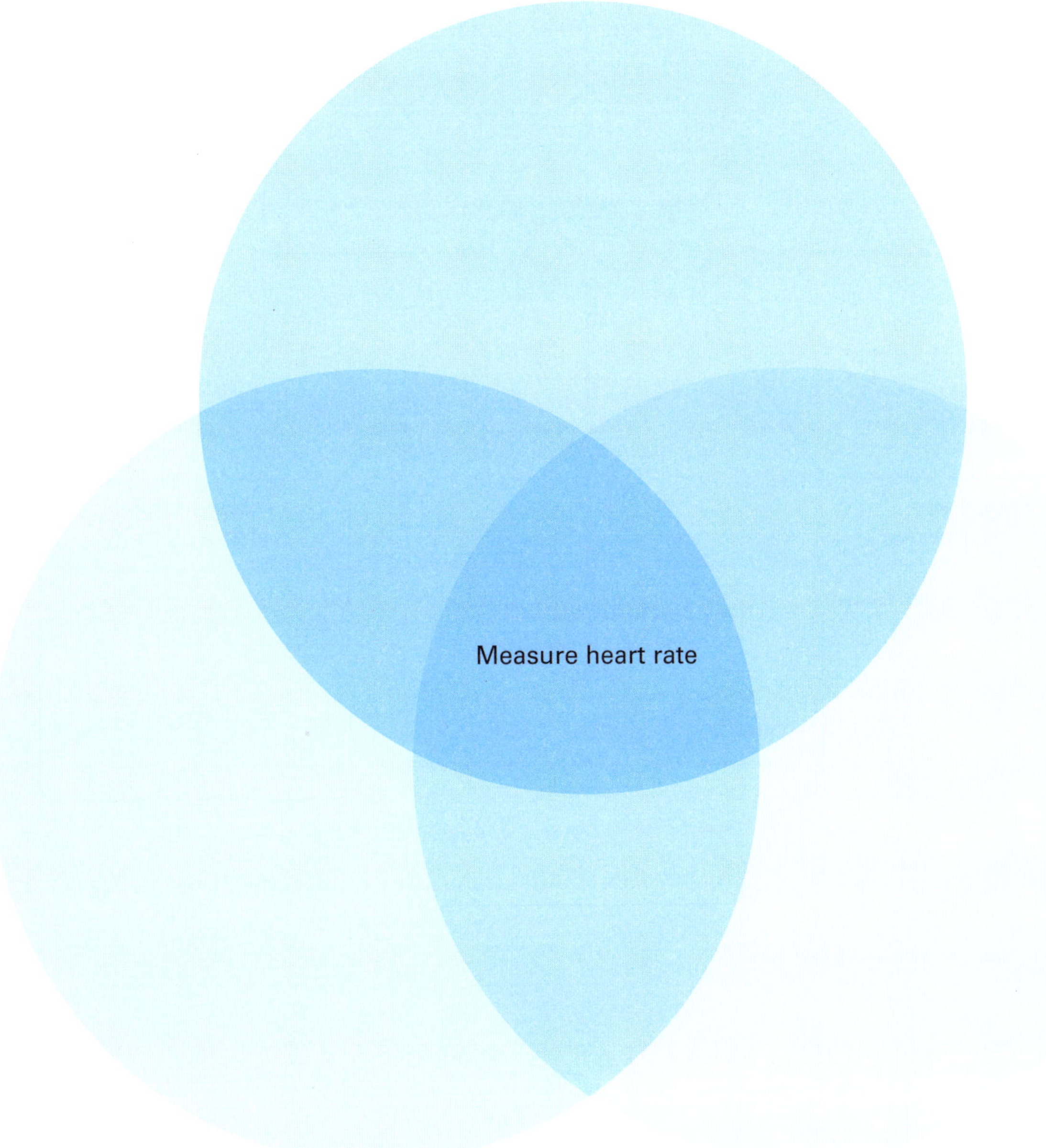

9780170465533

WORKSHEET 3.4 THE WELLNESS WHEEL

The 'Wellness' wheel shown below shows eight different strategies for developing overall wellness. In each of the eight boxes, list another strategy you could use. It could be something you do every day, or every now and then, to improve your physical, emotional, cognitive, spiritual and social health.

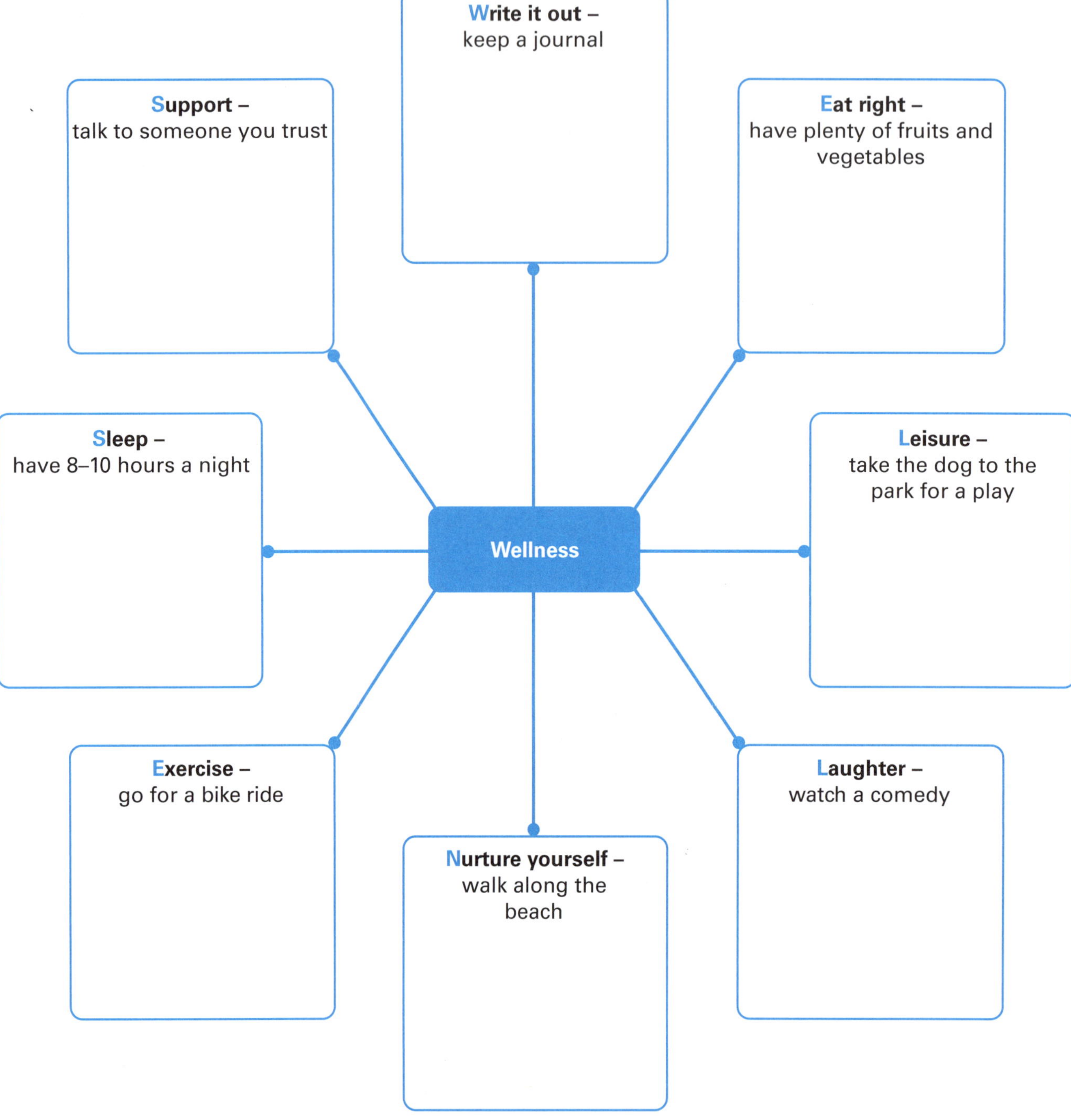

9780170465533

Investigation skills

WORKSHEET 3.5 CARDIOVASCULAR RESEARCH

Pages 119–20

Use the KWFL chart below to plan your research into cardiovascular disease. For each of the cardiovascular diseases listed, fill in each column of the table with:

- what you already **K**now
- what you **W**ant to know
- how you will **F**ind out
- what you have **L**earnt.

Topic	What you already **K**now	What you **W**ant to know	How you will **F**ind out	What you have **L**earnt
Arteriosclerosis				
Atherosclerosis				
Hypertension				

Once you have planned and conducted your research, present your information as a poster, brochure or web page.

WORKSHEET 3.6 REDUCING THE RISK OF CARDIOVASCULAR DISEASE

Page 120

Some risk factors are genetic and cannot be changed in a significant manner. However, many risk factors are both behavioural and environmental and can be directly influenced by making changes to either, or both, of these.

In the table below, profiles have been provided for four different people. You should suggest at least three strategies that each person could adopt to reduce their cardiovascular disease risk.

Person	Profile	Ways to reduce cardiovascular risk
Steph	Age: 22 Occupation: Uni student; part-time Uber driver Physical activity: Does not meet recommended guidelines but plays hockey once a week Diet: Mainly takeaway meals and 'food to go' from supermarkets due to little time to prepare meals Sleep: Averages 5–6 hours per night	
Faridah	Age: 16 Occupation: Student Physical activity: Exceeds recommended guidelines, but also spends about 3 hours per day gaming Diet: Does not eat recommended amounts of fruit and vegetables Sleep: Averages 8 hours per night	
Jake	Age: 72 Occupation: Retired Physical activity: Plays 6–8 hours golf on weekends Diet: High-fat, low-carb diet Sleep: Averages 6 hours per night	
Maz	Age: 34 Occupation: Administration Physical activity: Plays mixed netball once a week Diet: Skips breakfast regularly; eats microwave meals for lunch at their desk and often picks up takeaway on the way home Sleep: Likes to go out and often, and so only averages 4–5 hours sleep per night	

9780170465533

WORKSHEET 3.7 HOW 'HEALTHY' ARE YOU?

Are you as healthy as you think you are? Find out the state of your physical, social and cultural, emotional and spiritual health by taking this quiz. Record your responses for each section to find a total for each type of health, then add up all the numbers to get a final score.

	Rarely or never	Sometimes	Most of the time	Always
Physical health				
I get at least eight hours of sleep each night	1	2	3	4
I get colds, flu or infections	4	3	2	1
I get at least 30 minutes of walking, playing or other exercise every day	1	2	3	4
I eat fruits and vegetables every day	1	2	3	4
I do some form of relaxation and stress release daily	1	2	3	4
Social and cultural health				
I get along with the members of my family	1	2	3	4
I have someone that I can talk to about my feelings	1	2	3	4
I have a friend or friends of my own age group	1	2	3	4
I consider how what I say might affect others	1	2	3	4
I consider what I say before I speak	1	2	3	4
Emotional health				
I manage to 'bounce back' after difficult things happen in my life	1	2	3	4
I can express my feelings of happiness, sadness, fear or anger without feeling silly	1	2	3	4
I am an extreme worrier	4	3	2	1
I am able to adjust to changes in my life	1	2	3	4
When I am angry or annoyed with someone, I can let them know in a calm, respectful way	1	2	3	4
Spiritual health				
I believe that my life is worthwhile	1	2	3	4
I feel that I have my place in the world	1	2	3	4
I look forward to the things my life may offer now and in the future	1	2	3	4
I think that all people and living things have an important place on this earth	1	2	3	4
I believe that everyone has something to offer in life	1	2	3	4

Source: *Everybody's Different*, Jenny O'Dea, 2007. Reproduced by permission of the Australian Council for Educational Research.

WORKSHEET 3.7 CONTINUED

PHYSICAL HEALTH SCORE

- **13–17 points:** you have excellent physical health and you can add five more points if you don't smoke. A perfect score for you will be 22.
- **10–12 points:** you are moderately healthy and you can add five more points if you don't smoke. You need to take more care of yourself. Think of ways that you can look after your physical self.
- **5–9 points:** you are in relatively poor physical health but you can add five more points if you don't smoke. You need to work on all aspects of your physical health. Even small changes will help you to become healthier.

SOCIAL AND CULTURAL HEALTH SCORE

- **15–20 points:** you have excellent social health. You have people who care about you and support you, and you take care with the feelings of others. You might be a bit of a social butterfly!
- **10–14 points:** you are moderately healthy from a social point of view. You need to focus a little more on developing friendships and social relationships.
- **5–9 points:** you are in relatively poor social health. You need to work on all aspects of your social health, relationships and social networks.

EMOTIONAL HEALTH SCORE

- **13–17 points:** you have excellent emotional and mental health. You manage your emotions very well and you adjust to upsets and changes in your life very well.
- **10–12 points:** you are moderately healthy. You need to take more care of your emotional needs by expressing your feelings and talking things through with others.
- **5–9 points:** you are in relatively poor emotional health. You need to work on all aspects of your emotional health.

SPIRITUAL HEALTH SCORE

- **15–20 points:** you have excellent spiritual health.
- **10–14 points:** you are moderately healthy. But you need to take more care to recognise your spiritual beliefs and the meaning of your life.
- **5–9 points:** you are in relatively poor spiritual health. You need to work on all aspects of your spiritual health and the meaning and satisfaction of your life.

TOTAL HEALTH SCORE

- **61–79 points:** you have excellent overall health and score well on all the different health scales. Keep it up and you can look forward to a healthy life.
- **45–60 points:** you are moderately healthy, but you need to take more care of yourself in some areas. You may be doing quite well at some aspects of your health and neglecting other areas. Remember to take care of your 'whole' self and not just some aspects of your overall health.
- **22–44 points:** you need to work on all aspects of your health. Ask someone to help you to think of ways of becoming a healthier person.

Jenny O'Dea, *Everybody's Different*, ACER Press, 2007

WORKSHEET 3.8 COMPONENTS OF HEALTH

Summarise the benefits of regular physical activity on each component of health.

Page 128

summarise to give a brief statement of a general theme or major point(s)

Component of health	Benefit of regular physical activity
Physical (all body systems)	
Social (connectedness with others)	
Emotional (thoughts, feelings and behaviours)	
Cognitive (understanding and processing information)	
Spiritual (able to find hope and comfort)	

WORKSHEET 3.9 FACTORS FOR PARTICIPATION

Page 129

There are six factors identified in the cause and effect fishbone diagram below that can influence participation in physical activity. They are:

- culture
- peers
- family
- environment
- individual
- cost.

explain to make an idea or situation plain or clear by describing it in more detail or revealing relevant facts

Explain how each factor may cause a *decrease* in the physical activity levels of an individual.

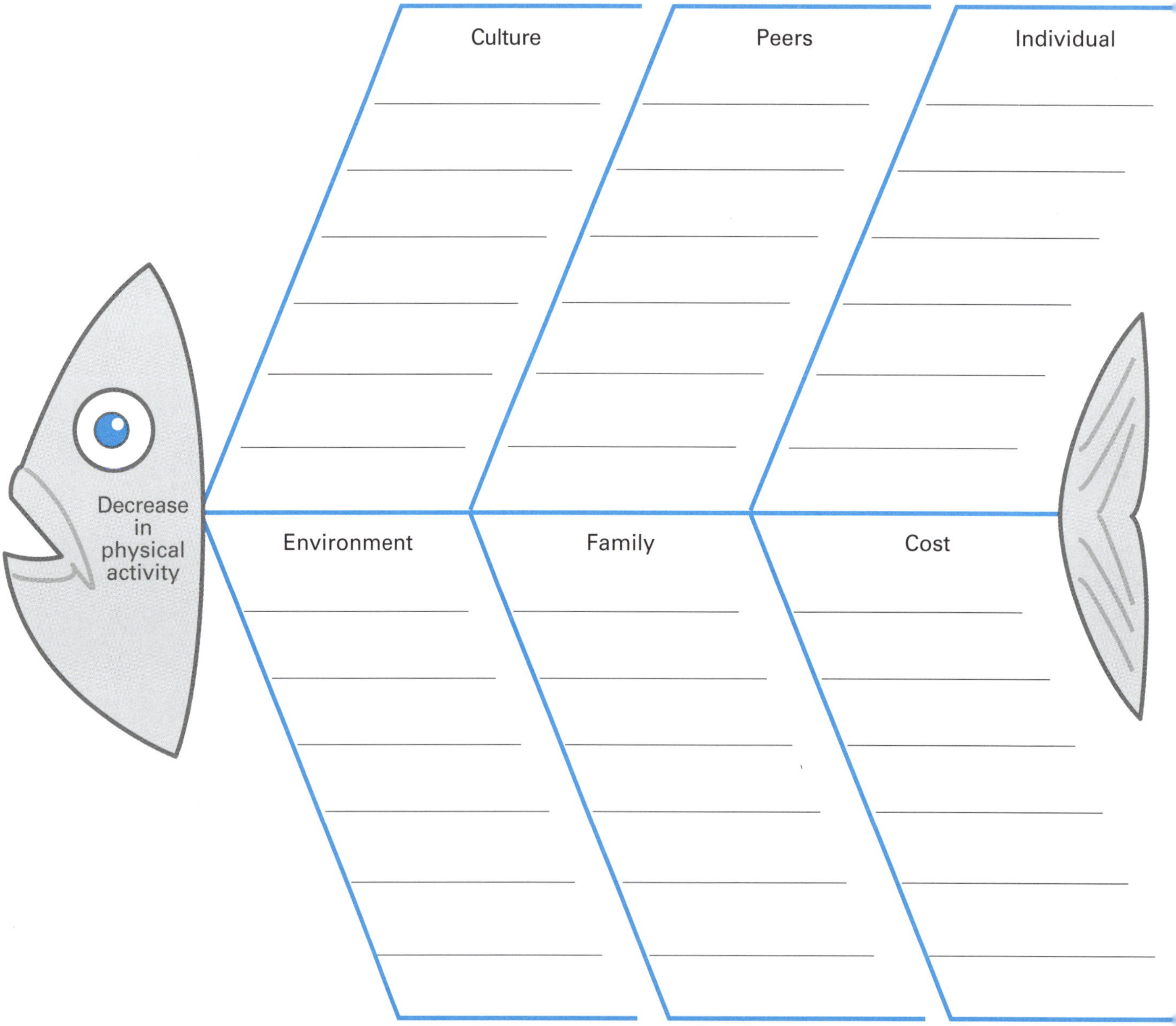

Decrease in physical activity

Culture

Peers

Individual

Environment

Family

Cost

WORKSHEET 3.10 BE A CRITICAL CREATIVE THINKER

Pages 134–7

Use the information in your student book about Australian participation in physical activity to help you complete the tasks in the following table. Choose one activity from each row to complete.

	Compare	Evaluate	Analyse
Row 1	Compare the proportion of children who meet both the physical activity and the sedentary behaviour guidelines at different ages. (Refer to Figure 3.18 in student book.)	Evaluate the proportion of Indigenous vs non-Indigenous children who meet the physical activity guidelines. (Refer to Figures 3.19 and 3.20 in student book.)	What happens to the proportion of Indigenous children who meet the activity guidelines as they become secondary age (12+)? (Refer to Figures 3.19 and 3.20 in the student book.)
	Perspectives	**Reflect**	**Imagine**
Row 2	What is the impact of the decrease in physical activity levels from the perspective of: • the Australian Government? • an individual? • the community?	Think about why the proportion of Australians engaged in sufficient physical activity decreases as we get older as adults (18+). Summarise your conclusions.	Imagine you are given the task of raising the physical activity levels of Australians who are 50+. How would you manage this?
	Classify	**Justify**	**Invent**
Row 3	Group each of the 'Top 20' activities listed in Figure 3.23 in the student book as individual or group sports. Comment on the participation rates for both groups.	The government has mandated that every school student in Years F–10 must do 30 minutes of structured physical activity during the school day. Justify this decision.	Create a tool (digital or non-digital) that helps individuals keep track of their sedentary behaviour levels.

Investigation skills

WORKSHEET 3.11 CLASS SURVEY

SB Pages 134–7

Survey your class to find out what physical activity, sport and recreation activities your classmates participate in. Use the activities listed below. Under 'other' write what your 'other' is; for example, bike riding, BMX racing or surfing.

Recreation activities	
Athletics, track and field	Martial arts
Australian Rules football	Netball
Basketball	Rugby League
Cricket (outdoor)	Rugby Union
Dancing	Soccer (outdoor)
Gymnastics	Swimming/diving
Hockey	Tennis
Horse riding/equestrian/polo	Other:

Count the number of responses for each activity and convert the total number in each sport to a percentage:

$$\frac{\text{Number of students who do the activity}}{\text{Number of students in the class}} \times 100 = \%$$

Graph your results on the axes below.

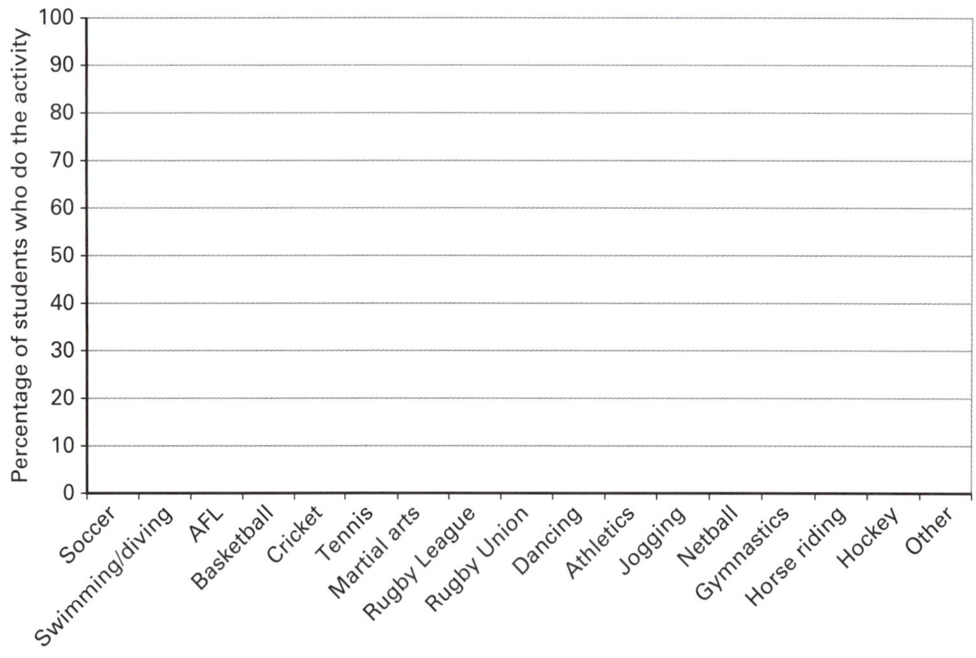

9780170465533

1 What is the most popular sport in your class?

2 What is the least popular sport in your class?

3 How do the class results **compare** with the Sport Australia data (refer to student book) about the most popular sports in Australia? Identify similarities and differences in the data.

compare to observe or note how things are similar or different.

AC

4 How can gender affect the sports and activities that you participate in? Use the class data and Sport Australia data to support your answer.

9780170465533

5 Suggest three reasons why athletics has one of the highest participation rates in Australia.

6 How can the area where you live influence or affect the sports and activities that you participate in? Use the class data to support your answer.

WORKSHEET 3.12 ENVIRONMENT AND PHYSICAL ACTIVITY

SB
Pages 134–7

Use Google Earth to obtain a satellite image of an area within your community you know provides multiple opportunities for people to engage in physical activity. This could be walking tracks, parks, tennis courts, indoor facilities, pools, etc.

Here is an example:

Imagery © 2020 CNES/Airbus, Maxar Technologies, Map data © 2020

Complete the following table by summarising all the potential activities people could engage in if they were to be 'dropped' into the location you have chosen.

Environmental feature/facility	Potential for physical activities

WORKSHEET 3.13 KEY TERMS

Some key terms about physical activity are listed in the first column of the table below. Your task is to fill out the second column with your own definition of the word, the third column with the *actual* definition and fourth column with a sentence that uses this word in the context of this chapter.

Key terms	What I think it means	What the student book or dictionary says it means	How I can use it in a sentence about health and physical activity
Physical activity			
Sedentary behaviour			
Intensity			
Frequency			
Energy expenditure			
Emotional health			
Social health			
Cognitive health			
Wellbeing			
Resilience			

9780170465533

CHAPTER 3 REVIEW

Reflect upon your learning in this chapter. Completing the following sentences will help to structure your thoughts about the health benefits of physical activity.

1 I think we learnt about this topic because:

____

____

2 I wondered:

____

____

3 I thought:

____

____

4 I now know:

____

____

5 I was surprised that:

____

____

6 I will always remember:

____

____

7 I still wonder:

____

____

8 I now plan to:

____

____

4

MENTAL HEALTH AND WELLNESS

9780170465533

WORKSHEET 4.1 WHAT AFFECTS YOUR MENTAL HEALTH?

List two things you could do to improve each of your dimensions of health (physical, social, emotional, cognitive and spiritual) that will also have a direct impact on your mental health. For example, doing some exercise (physical) or hanging out with your friends (social) might help you feel better about yourself. Explain how each improvement will benefit your mental health.

Pages 146–50

Dimension of health	Improvement 1	Improvement 2
Physical		
Social		
Emotional		
Cognitive		
Spiritual		

WORKSHEET 4.2 WHO ARE YOU?

1 Create a list of all the positive things that make up YOU. Try to think of about 20 things and fill in each of the thought bubbles below with five things.

2 In the thought bubble at the bottom of the page, write a paragraph about yourself using the qualities you have identified. Write as if you are describing yourself to a relative whom you have never met.

Things I do well

Things I like to do

Getty Images/Jack Hollingsworth

Things that make me happy

The way I look

3 Using the lists you have made, think about whether you would be the same person if you:

- were born in a different country
- lived in a different part of Australia
- were another gender
- had a different skin colour
- were taller/shorter
- had a different ability/disability.

Choose three of these points of difference and write a sentence about the ways in which you might be different. These could be your attitudes, traditions, cultural practices, educational opportunities, sporting ability, friendships, daily life, body image, and so on.

a ______________________________

b ______________________________

c ______________________________

WORKSHEET 4.3 WHAT'S THE DIFFERENCE?

1 Conduct an online search for information about photo-retouching on social media and then answer the following questions.

a What is photo-retouching?

b **Propose** why people retouch their photos before posting them on social media?

Propose to put forward (e.g. a point of view, idea, argument, suggestion) for consideration or action

AC

c Do you believe photo-retouching should be allowed? Why/why not?

d What impact does seeing retouched photos on social media have on young people?

e Do you think that most young people know about photo-retouching?

f **Consider** if your research changes the way that you think about the media?

Consider to put forward (e.g. a point of view, idea, argument, suggestion) for consideration or action

AC

g What would you say to the media industry about this practice?

2 Write a paragraph either for or against digital alteration of your annual school photos. Introduce the paragraph by stating your opinion, whether you agree or disagree with the use of digital alteration. Identify two or three reasons why you have this opinion and give examples or show the impact of these reasons. Finish by summing up your decision.

WORKSHEET 4.4 STRESS LESS

Pages 157–61

Make a list of the stressors you have in your life and then identify how you know you are stressed. An example has been done for you.

Stressors	How you know that you are stressed	Coping strategies
Example: fighting with a friend	Not able to concentrate on my school work Not sleeping properly	Going to talk to my friend to resolve the fight

WORKSHEET 4.5 SELF-TALK

Pages 163–4

Negative self-talk can be very destructive – so much so that we can start to believe it. So, we need to be mindful and stop ourselves when we hear it starting. Think about the types of negative self-talk that you might say, if any. The following are some common negative self-talk statements; see if you can come up with two different responses to each of them.

Instead of:	**Say:**
'I'm not good enough.'	Example: I'm going to try my best. ____________________ ____________________
'I'm too slow.'	Example: I can practise and get faster. ____________________ ____________________
'I'm a loser.'	____________________ ____________________
'They won't like me.'	____________________ ____________________
'I'm going to fail.'	____________________ ____________________
'I'm dumb, I can't.'	____________________ ____________________
'I'm not thin enough.'	____________________ ____________________

WORKSHEET 4.6 LISTENING QUIZ

Pages 164–6

How good are you at listening? For each statement, rate how often you do this using the scale below:

4 = Always **3** = Most of the time **2** = Some of the time **1** = Rarely

Statement		Rating
1	I remind myself that listening is an opportunity to learn something.	
2	I make myself listen even when the subject fails to interest me.	
3	I make myself listen even if I fear that the speaker is a moron.	
4	I focus on the speaker's ideas and not on appearance or mannerisms.	
5	I try to understand the speaker's feelings as well as his or her words.	
6	I wait for the speaker to finish before forming an opinion.	
7	I make sure I understand the speaker's point of view before responding.	
8	I show that I am engaged by maintaining eye contact, nodding, and leaning forward.	
9	I am relaxed, calm, and patient when listening.	
10	I do not interrupt the speaker.	
11	I take notes if I must remember points being made.	
12	I capitalise on lag time by reviewing in my mind the speaker's main ideas.	
13	I ask nonthreatening questions to ensure that I understand.	
14	I often paraphrase what I hear to be sure I have heard it correctly.	
15	I do not allow distractions to divert my attention.	

Score:

50–60 Excellent! You have exceptional listening skills.

40–49 Above average, but you could improve your skills.

30–39 Your score is promising, but you could greatly improve.

15–29 You could improve your listening skills. Perhaps ask for help?

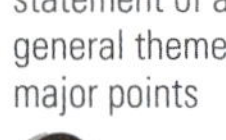

Summarise to give a brief statement of a general theme or major points

1 How did you go? Are you a good listener? **Summarise** your results.

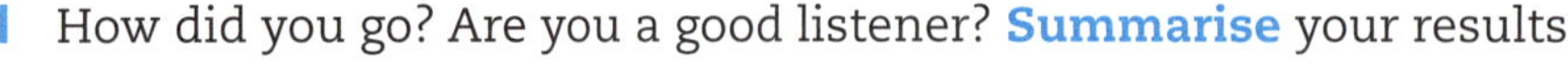

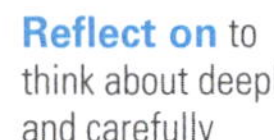

Reflect on to think about deeply and carefully

2 **Reflect on** any ways that you could improve your listening.

9780170465533

WORKSHEET 4.7 REAL WORLD CASE STUDY

Read the case study below, then answer the questions.

CASE STUDY

Ahn is a friend of yours at school; you don't really hang out with him every day, but he is in a few of your classes. Lately he hasn't been himself and you can't really figure out what it is. He just looks sad all the time. No one else seems to notice and he seems to be spending more and more time at school by himself. He has been telling everyone that he goes to the music room at lunchtime to practise, but you saw him behind the sports storeroom, sitting alone. You feel concerned for him and decide to see what is wrong. You sit down with him and he tells you that his brain feels all 'fuzzy' and he can't concentrate He thinks he is going to fail all his subjects and he is really worried.

1 Consider what you might do immediately.

2 Discuss what your actions might be over the next few days.

3 Who might be able to help Ahn? What could they do?

WORKSHEET 4.8 RESPECTING DIVERSITY

In this activity, you will look at ways in which cultural diversity can be respected.

In each petal of the flower diagram below, identify an area in which your school can show that cultural diversity is valued and promoted. The first petal has one area of school life identified for you (Canteen menu). Have a look at your school's canteen menu. Are there items on the menu that show consideration of cultural diversity, such as sushi, curry or noodles? Another example may be in your PE classes, when you play games that originated in other countries. What other areas can you identify?

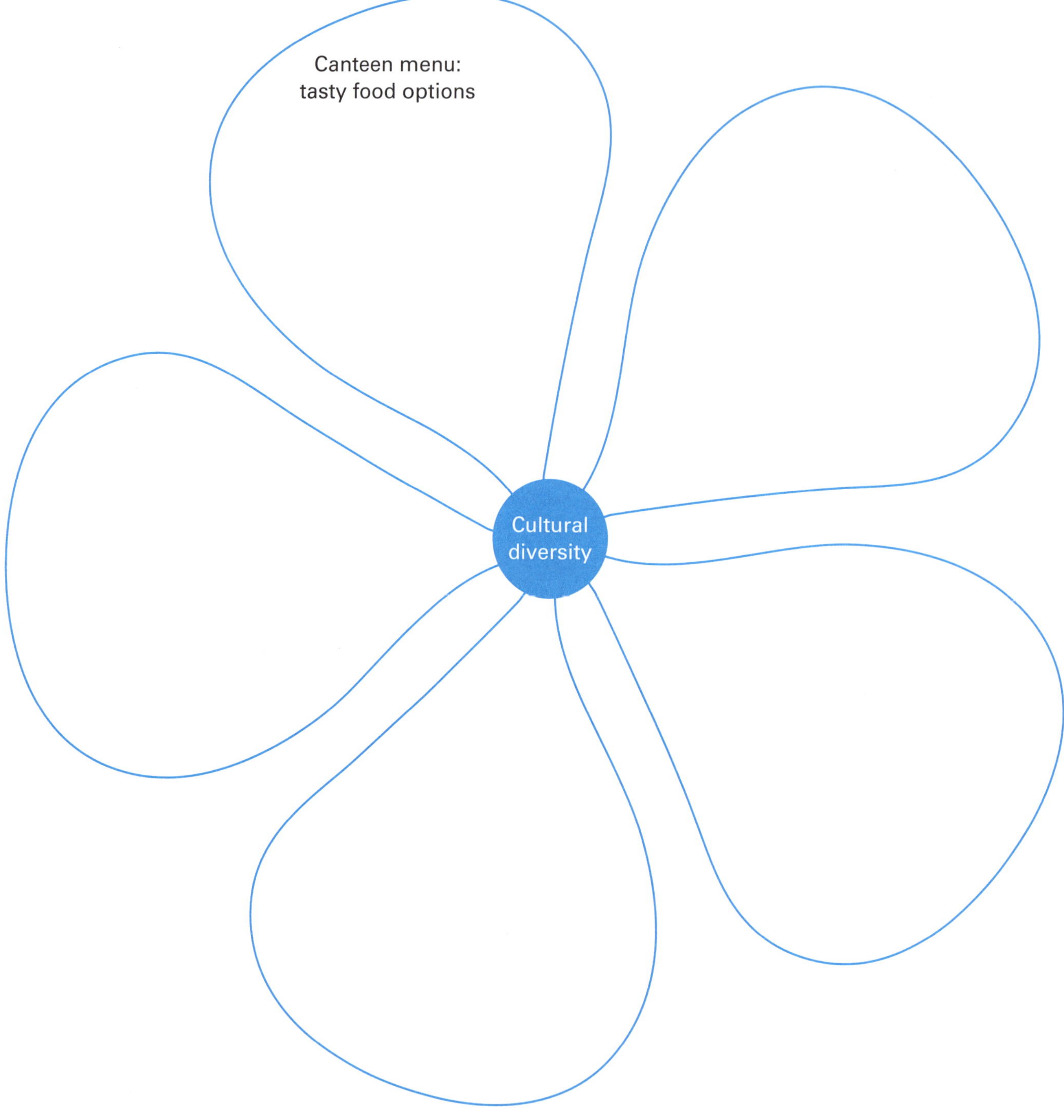

9780170465533

WORKSHEET 4.9 HOW CAN WE FIT RELAXATION INTO OUR LIVES?

Think about how you spend your days. Fill in the timetable below and see where you could fit in some time to relax. Write in some of the things you like to do to relax. Maybe if you schedule them you will be more likely to do them!

SB Pages 174–6

	Monday	Tuesday	Wednesday	Thursday	Friday	Saturday	Sunday
6–8 a.m.							
8–10 a.m.							
10–12 a.m.							
12–2 p.m.							
2–4 p.m.							
4–6 p.m.							
6–8 p.m.							
8–10 p.m.							

WORKSHEET 4.10 GETTING HELP

Pages 176–8

Place the following sources of mental health information in the appropriate circle on the diagram below. You need to decide whether the source is most reliable, possibly reliable or not reliable.

Family	Counsellors	Kids Helpline
Friends	Doctors	Lifeline website
Relatives	Nurses	Beyond Blue website
Teachers	Chemists	Wikipedia
Shopkeepers	Newspapers	Advertisements
Magazines	Police	Facebook
Sports coaches	TikTok	TV shows
Brochures	Butterfly Foundation website	The government

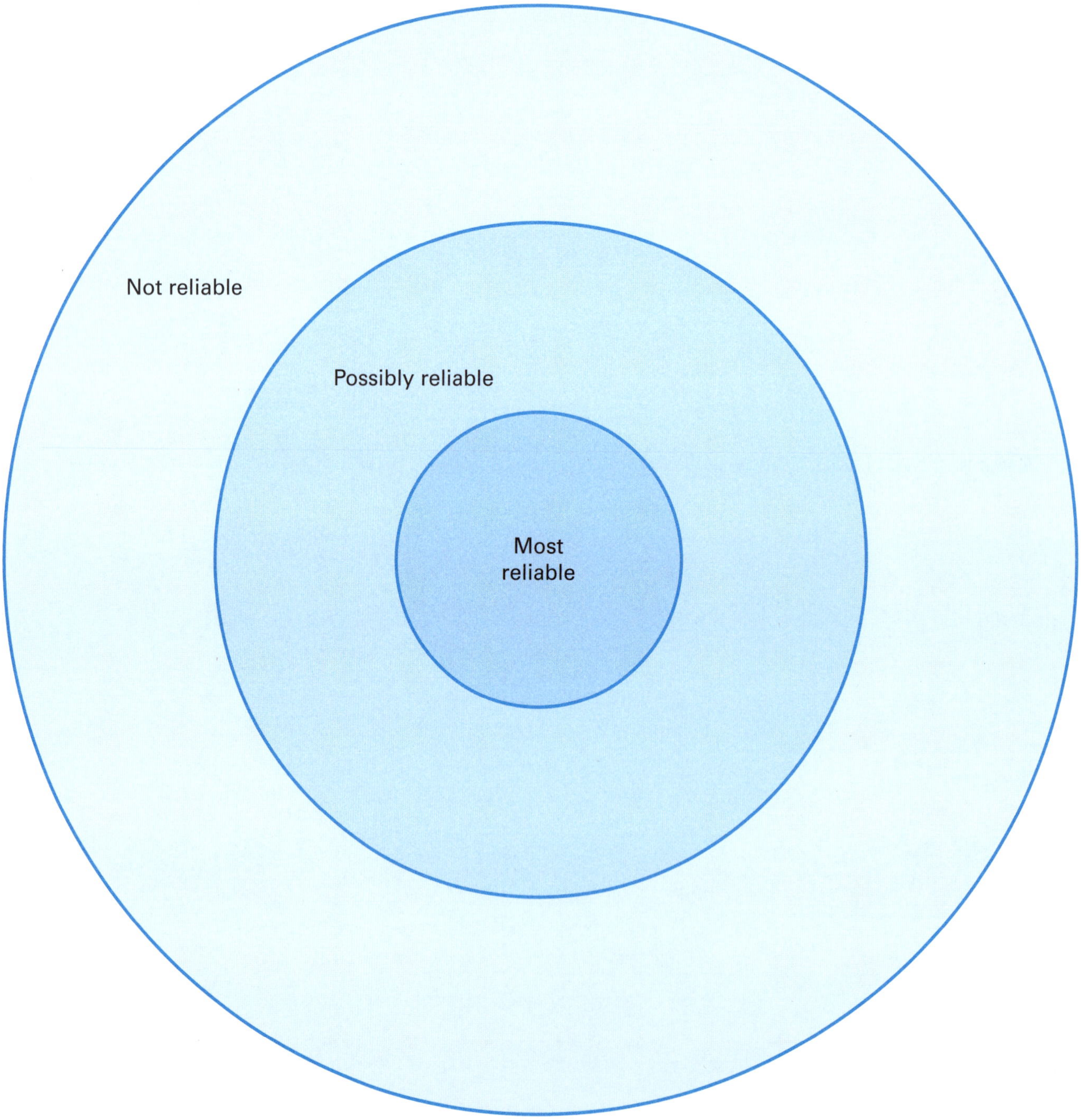

Not reliable

Possibly reliable

Most reliable

9780170465533

WORKSHEET 4.11 DO YOU SEE WHAT I SEE?

Page 180

Check out the 10 October World Mental Health Day clip 'Do you see what I see?' on YouTube. Think about what we have talked about in the chapter so far when you are answering the questions. Use your new knowledge, understandings and skills to complete this activity.

Weblink 'Do you see what I see?' World Mental Health Day, 10 October

1 Discuss how this clip makes you feel.

2 Discuss how you would describe the clip to a friend.

3 Why is there a black/white side and a colour side in the drawing?

4 Summarise the top three messages for viewers. (Remembering it has been made for World Mental Health Day.)

5 Consider any other messages that could be included in the clip. What might these be?

____

____

____

____

____

Identify to distinguish; locate, recognise and name; provide an answer from a number of possibilities

6 **Identify** any techniques that have been used to get the messages across, considering they haven't used any speaking words.

____

____

____

____

7 Show this clip to a family member or friend, ask them for their thoughts on the clip; for example, do they think it is effective in getting a message across. Refer back to the messages you got out of watching the clip.

____

____

____

____

8 Consider whether there is anything you would change to make it more relevant/reliable/effective.

____

____

____

____

Weblink How to make a storyboard

9 Design your own storyboard for a clip that would promote the messages you think are the most important. Limit your target group just to secondary school students. You might want to develop and film this clip to show in your local health promotion activity.

WORKSHEET 4.12 PROTECTIVE AND RISK FACTORS

Pages 179–80

The protective and risk factors for mental health and wellbeing below come from the Melbourne Charter produced by VicHealth. When protective factors are put in place for us, we will have a better chance of having good mental health and wellbeing.

Protective factors for mental health and wellbeing:

- Arts and cultural engagement
- Childhood: positive early childhood experiences, maternal attachment
- Cultural identity
- Diversity: welcomed/shared values
- Accessible education
- Safe environments
- Empathy
- Empowerment and self-determination
- Family: resilience, parenting competence, positive relationship with family members
- Affordable, fresh food
- Affordable, accessible housing
- A reliable income, with safe accessible employment and work conditions
- Personal resilience and social skills
- Physical health
- Respect
- Social participation: supportive relationships, involvement in group and community activity and networks
- Participation and access to sport and recreation
- Transport: accessible and affordable
- Accessible quality health and social services
- Spirituality

Risk factors for mental health and wellbeing:

- Alcohol and drugs: access and abuse
- Disadvantage: social and economic
- Displacement: refugee and asylum-seeker status
- Disability
- Discrimination and stigma
- Education: lack of access
- Environments: unsafe, overcrowded, poorly resourced
- Family: fragmentation, dysfunction and child neglect, post-natal depression
- Food: inadequate and inaccessible
- Genetics
- Homelessness
- Isolation and exclusion: social and geographic

- Natural and human-made disasters
- Rejection by classmates
- Political repression
- Physical illness
- Physical inactivity
- Poverty: social and economic
- Racism
- Poor employment conditions and insecure employment
- Violence: interpersonal, intimate and collective; war and torture
- Work: stress and strain

The Melbourne Charter for Promoting Mental Health and Preventing Mental and Behavioural Disorders, VicHealth, 2009

1 Choose three protective factors from the list above. Suggest a way that your school could support or do something to address each of these factors. For example, regarding the protective factor of affordable and quality food, the school canteen could make sure there is healthy, fresh food available to students at a reasonable price.

Protective factor 1

Protective factor 2

Protective factor 3

2 Now choose three risk factors and discuss how the school could help overcome these risks. For example, regarding the risk factor of inadequate food, the school could start/continue a breakfast program to make sure all students have access to breakfast.

Risk factor 1

Risk factor 2

Risk factor 3

3 Identify other organisations in your community that help promote protective factors for young people and describe how they do it.

4 Consider what organisations are available to assist parents provide a positive environment for their children.

WORKSHEET 4.13 SOCIAL ISOLATION

Page 181

One of the protective factors is social participation. This is important in preventing social isolation and the mental health issues this brings.

Think about what it might be like to live in a small country town if you live in a city, or what it might be like to live in a city if you live in a small country town (so think about the opposite of where you live).

1 What groups, activities or organisations would you advise young people who have recently left school to engage in to promote social participation?

2 What groups, activities or organisations would you advise newly retired people to engage in to promote social participation?

3 **Contrast** any differences that are present in cities or small towns.

Contrast to display recognition of differences by deliberate juxtaposition of contrary elements; show how things are different or opposite

AC

4 Consider and discuss whether it would be easier or harder to promote social participation in a city rather than in a small town.

9780170465533

This time think about someone who is 18, they have just finished Year 12 and have moved to a new town/city for university or work. They don't know anyone in the town/city and their closest friends live 100 km away.

5 Choose three of the protective factors (not education, income or social participation) and. **Discuss** how participating in these could improve the 18-year-old's mental health and wellbeing. For example: attending an Arts event or enrolling in a class might make the person happy, want to talk to others there, share their own creativity, make friends with people who are like-minded, get out of the house, have something to talk about with others at work/university/home, feel like they are becoming part of a community, improve a skill, etc.

discuss to talk or write about a topic, taking into account different issues or ideas

AC

a

b

c

6 Consider what this person might have to do if they were very shy. If you were their friend who lived far away, what would you recommend them doing to combat this shyness?

7 Outline three services or community groups your friend could get in touch with to help them 'settle in' to their new town/city?

WORKSHEET 4.14 USING CREATIVITY TO PROMOTE MENTAL HEALTH

Using the knowledge, understanding and skills you have learnt across this whole chapter, choose one of the following and research this art form to complete the task:

Weblink
Dreamtime stories

1 In Aboriginal and Torres Strait Islander cultures, storytelling is a component of teaching and learning. Write a short story book for primary-school-aged Aboriginal and Torres Strait Islander children to teach them about good mental health.

2 When you think of India, you might think of Bollywood. The films they produce are largely musicals, which use catchy lyrics and dance moves. Write a song that promotes mental health and come up with some dance moves that you could use to help you get the message across.

3 In some countries in the Caribbean, Central America and South America people celebrate Carnival once a year, where they dress in elaborate costumes and parade down the streets to local music. Design a costume that represents positive mental health and discuss how you might be able to use it in Australia.

Alamy Stock Photo/Bill Bachman

Alamy Stock Photo/Matt Crossick

CHAPTER 4 REVIEW

Reflect upon your learning in this chapter. Completing the following sentences will help to structure your thoughts about mental health and wellness.

1 Three support strategies I can access when I am feeling down are:

2 Three ways I can support a friend or family member who is struggling are:

3 I think we learnt about mental health because:

4 I will always remember:

5 I now plan to:

6 I would like to learn more about:

5

PUBERTY AND RESPECTFUL RELATIONSHIPS

9780170465533

WORKSHEET 5.1 VENN DIAGRAM

Pages 186–94

Come up with a list of the physical changes that happen at puberty. Identify which changes happen to most people with testes, which happen to most people with ovaries and which happen to both and fill them in to the Venn diagram below.

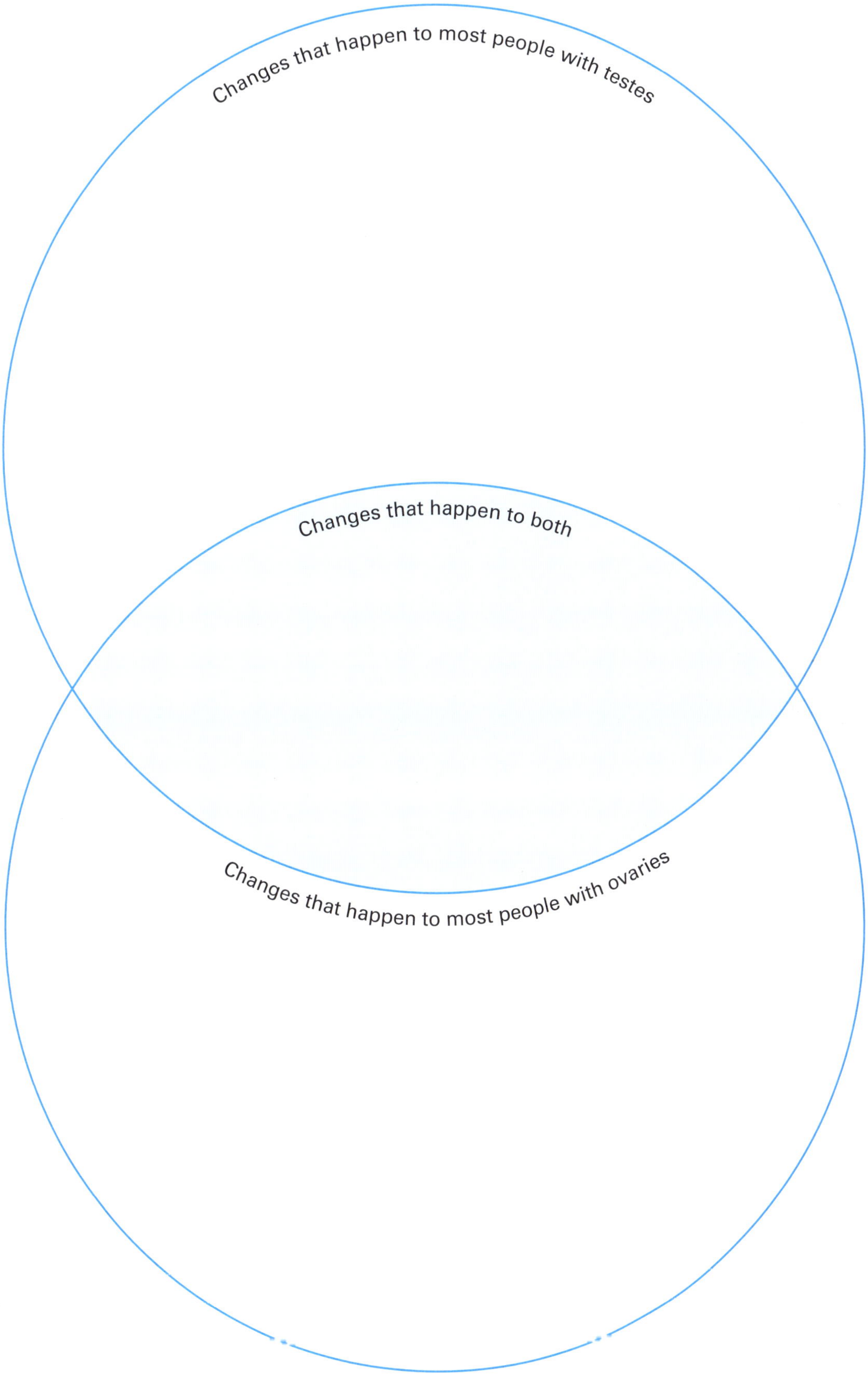

Changes that happen to most people with testes

Changes that happen to both

Changes that happen to most people with ovaries

WORKSHEET 5.2 LIFE STAGES POSTER

Using the age guidelines in the table below, design a poster that demonstrates the characteristics of one life stage. Your teacher will suggest a particular stage, so all the stages will be covered by the whole class. Use images from magazines, online or draw your own pictures.

Think about the following when you are designing your poster:

- What signifies the start of this stage? For example, the start of the first stage is the sperm meeting the egg.
- What signifies the end of this stage?
- What physical changes occur during this stage?
- What can they do now that they couldn't do in the previous stage?
- What expectations are placed on them in this stage?
- What types of things will shape their identity?

Stages of life	
Conception to birth	40 weeks
Infancy	0–2 years
Childhood	3–12 years
Puberty/adolescence	13–18 years
Adulthood	19–39 years
Middle age	40–65 years
Old age	65 years+

WORKSHEET 5.3 DEAR Dr

You have a regular column in a teen magazine where you offer support for young people in regards to the changes occuring. You have received these three letters and want to reply in the magazine because you think a lot of young people might have the same questions.

Dear Dr,

All my friends have their period and I'm the only one missing out. I am 15 and thought everyone should have it by now. What is wrong with me? Is there any way I can hurry it up?

Thanks,

Concerned

Dear Concerned,

____

____

____

____

____

____

Dear Dr,

My brother has turned into such an idiot. I mean, he never used to be one. But since puberty has hit, he is not the same. He seems so angry all the time, He slams doors and yells at Mum and Dad. Tbh puberty hasn't treated him well; he's got bad acne and smells awful. Is this why he is so annoying?

From,

Brotherly love

Dear Brotherly love,

____

____

____

____

____

____

Dear Dr,

Please help, I don't feel like me anymore. Ever since puberty hit, and it hit me early and hard, everyone expects so much more from me. I still feel like a kid inside, but now have to think about adult stuff. I even got asked to go out with someone!! I'm not ready for that. What do I do?

From,

Growing up too fast

Dear Growing up too fast,

____

____

____

____

____

____

WORKSHEET 5.4 SPIDER DIAGRAM

On the spider diagram below, identify some ways your family or guardians have satisfied your physical, social, emotional and cultural needs. See if you can come up with at least six examples for each of the needs.

SB Pages 200–6

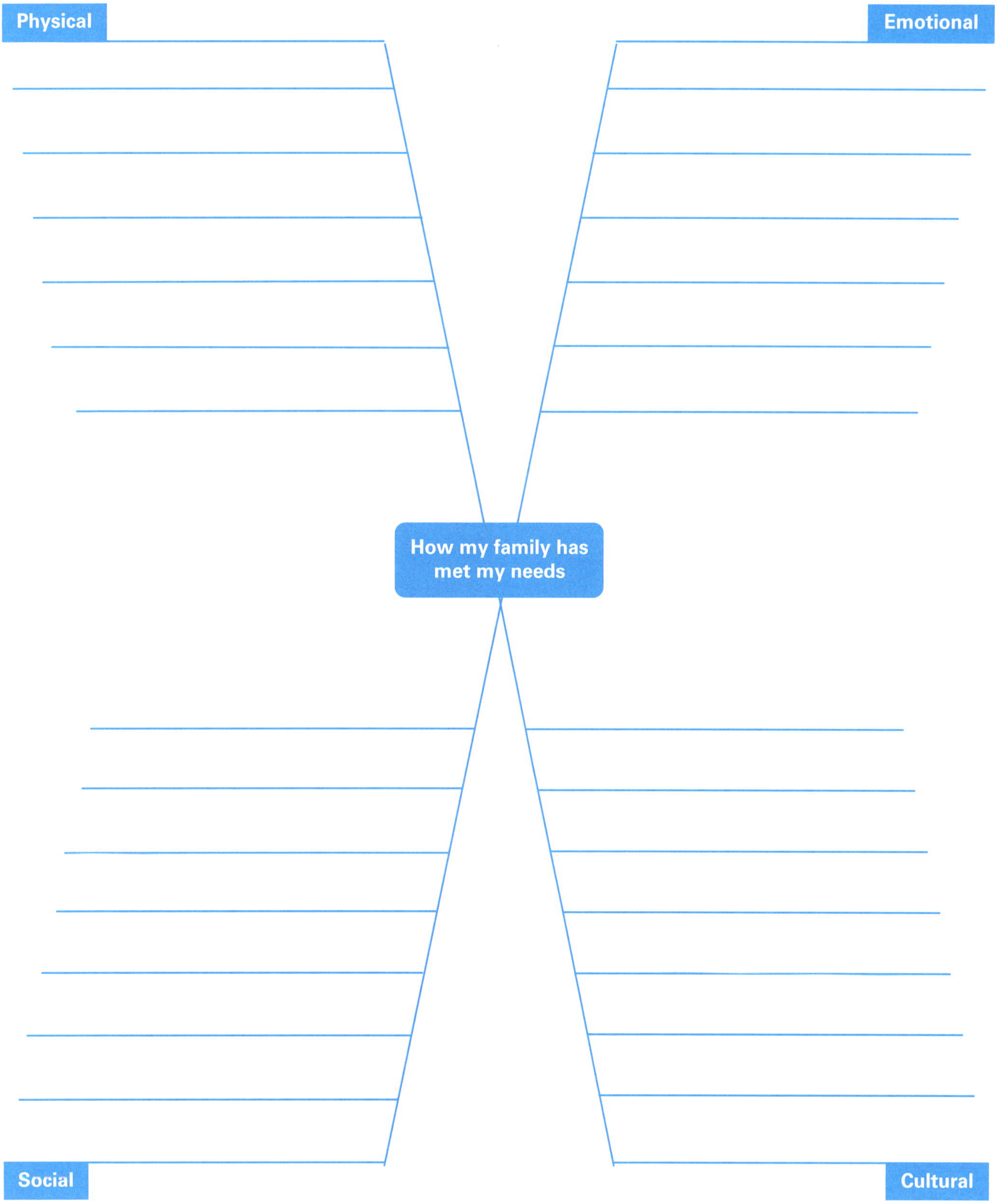

Physical

Emotional

How my family has met my needs

Social

Cultural

WORKSHEET 5.5 FAMILY TYPES

Pages 200–1

CASE STUDY 1

The Chu family migrated to Australia when Min was eight and his sisters were six and four years old. Min's parents worked hard to save some money in order for Min's paternal grandmother to come and live in Australia with them. They bought a small grocery store and his grandmother looks after the children before and after school when their parents are working. All three generations live in the same house.

Alamy Stock Photo/Asia Images Group Pte Ltd

CASE STUDY 2

Lucinda is 14 and she has a 10-year-old sister, Sarah. Lucinda's parents have divorced, and she and her sister live with their mother, who works full-time to support the family.

iStock.com/Alex Potemkin

CASE STUDY 3

Sam and Alex got married not long after the marriage equality laws came in; they had been together for 12 years previously. They have since adopted two children, Sierra (6) and Ben (6 months). Sam works full-time for a prestigious law firm in the city and Alex is a writer, who can usually work from home.

iStock.com/monkeybusinessimages

WORKSHEET 5.5 CONTINUED

5

1 **Identify** the type of family that is represented in each scenario.

identify to distinguish; locate, recognise and name

AC

Case study 1

Case study 2

Case study 3

2 **Consider** how the responsibilities might be shared among family members in each family type. Give three examples for each case study.

Consider to put forward (e.g. a point of view, idea, argument, suggestion) for consideration or action

AC

Case study 1

Case study 2

Case study 3

WORKSHEET 5.5 CONTINUED

3 Complete the PMI (Plus, Minus and Interesting) table for the family types above. Describe one or two advantages and disadvantages of each family type, and one interesting point about each family type.

	Plus	Minus	Interesting
Family type 1			
Family type 2			
Family type 3			

WORKSHEET 5.6 THE RELATION SHIP

1 Read the qualities listed below. Choose the four that are most important in your relationship with your family, and the four that are the most important qualities of your friends. Write them into the sails of the relation ship in your priority order.

Love	Humour	Support	Problem solving
Trust	Understanding	Openness	Praise
Communication	Loyalty	Laughter	Honesty
Encouragement	Patience	Compassion	Forgiveness
Respect	Generosity	Listening ability	Reliability

Family

Friends

2 One at a time, discuss with the person next to you which qualities you feel are most important and why. As a pair, come up with a priority order for the top four qualities for each type of relationship. Be prepared to justify your priority order reasoning to the class on a continuum on the board. Try to use real examples.

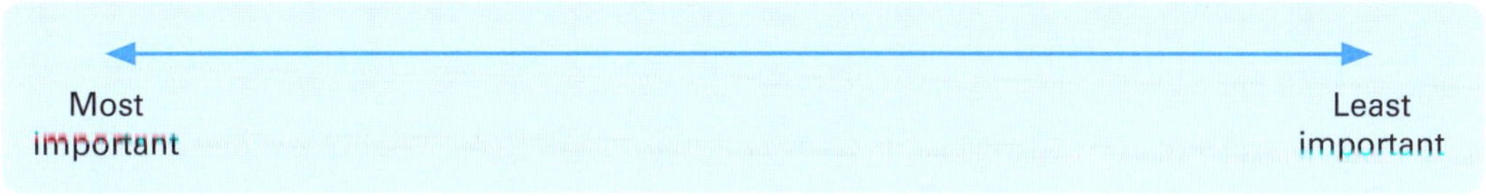

WORKSHEET 5.7 PEER PRESSURE

Pages 206–9

1 Write a script for a peer-pressure-related situation that a young person might find themselves in. Write two alternative endings for your script:

a one where the young person isn't assertive and gives in to peer pressure

b the other that demonstrates a strategy the young person could use to diffuse the pressure situation.

Script

Ending a

Ending b

Discuss to talk or write about a topic, including a range of arguments, factors or hypotheses

2 **Discuss** why someone might give in to peer pressure.

9780170465533

WORKSHEET 5.8 FRIENDS ONLINE

SB Pages 208–9

Using websites such as the eSafetyCommissioner site (an Australian Government site for help on anything online), research how you can stay safe online.

1 List five strategies we can implement to protect ourselves online.

2 Consider why companies want to get all the information they can about us. What do they do with that information?

3 Define what going 'incognito' means. How do we do this?

4 Draft a quick email/letter to your grandfather who doesn't understand all this 'internet stuff'. Explain why he should go 'incognito'.

5 Sometimes we think we are safer online, as we can take down something that was said, or a picture we don't like, or stay anonymous. What would you do in the following situations?

a Your friend asks for your password for Instagram because he wants to have a look at what someone said about him and he doesn't have access to see. What do you do?

____

____

____

____

b Your partner has asked you to send a naked picture of yourself to them and they promise that no one else will see it. What do you do?

____

____

____

____

c You have been talking to someone online for about two months now and you really like how they seem to 'get' you. They have been really polite and haven't pushed you to do anything. They are now asking you to meet up. What do you do?

____

____

____

____

d You see some bullying happening online and one of your friends wants you to make a comment to back them up. What do you do?

____

____

____

____

WORKSHEET 5.9 THINK, PAIR, SHARE: WHERE DO YOU SIT?

Page 210

Think:

In this activity you are to consider each of the statements and place a 'x' on the continuum where you 'sit'. For example, if you strongly agree you would place your 'x' near the far left of the continuum. Try to make a decision and stay either side of the unsure (middle). Jot down any notes you think of as to why you believe this, you will use these in discussions later.

1 Young people should be taught about respectful relationships at school.

Strongly Agree	Agree	Unsure	Disagree	Strongly Disagree

____

____

2 A partner wanting to be with you all the time is a sign that they love you.

Strongly Agree	Agree	Unsure	Disagree	Strongly Disagree

____

____

3 An adult always has more power than a young person.

Strongly Agree	Agree	Unsure	Disagree	Strongly Disagree

____

____

4 Posting about my relationship online tells everyone how happy I am.

Strongly Agree	Agree	Unsure	Disagree	Strongly Disagree

____

____

Pair:

Find a partner and discuss your answers: were they similar? Different? Why might this be the case?

Share:

In a large group your teacher might ask for volunteers to discuss their opinions or put your 'x' on a continuum on the board. Be prepared to discuss why you think the way you do

WORKSHEET 5.10 BREAK THE SILENCE

Weblink
Break The Silence – IHHP, White Ribbon, Ngukurr

1 Access and watch the weblink Break The Silence – IHHP, White Ribbon, Ngukurr (https://www.youtube.com/watch?v=9VDqW8vceeU).

2 As an example of a positive way to influence communities on respectful relationships, this clip promotes many messages. Fill in the wheel below describing what messages you heard directed to the different groups to break the silence.

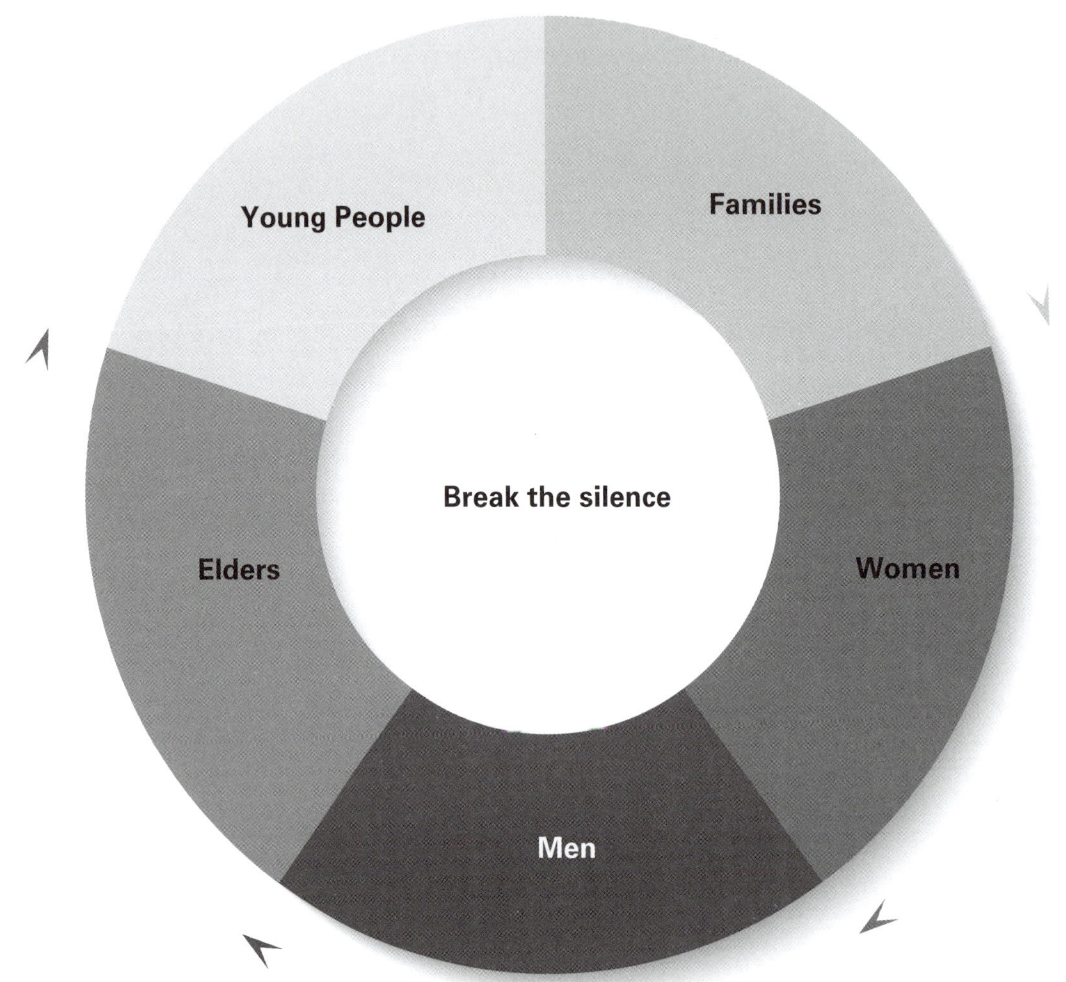

Shutterstock.com/Fireofheart

3 Describe the ways that this clip promotes inclusion in our communities.

____

____

____

4 Describe the ways that this clip values diversity in our communities

____

____

____

9780170465533

WORKSHEET 5.11 CONSENT

Consent means freely agreeing to something. Informed consent means you understand what you are giving consent for and there is nothing that will interfere with that decision (such as drug use, age, power in relationships, etc). Affirmative consent means that it is informed, explicit and voluntary, and that steps have been taken to gain consent.

Shutterstock.com/Sergio Vas

1 Discuss the following four scenarios as to whether consent was gained or given.

A Thom really likes Keira and keeps finding ways to see her during the day at school. He walks past her locker as often as he can. Kiera kind of likes the attention, but she is not ready for a relationship. Thom posts selfies on his social media pages and they always have Keira in the background.

	Yes	No
Thom gained consent to post the pictures		
Kiera gave consent to post the pictures		

B Archie and Lela have been going out for three weeks. Archie told Lela he was happy to kiss, but Archie doesn't enjoy Lela putting her tongue in his mouth. Archie feels bad and tries to avoid being alone with Lela. Lela attempts to kiss Archie in front of others because they are never alone.

	Yes	No
Lela gained consent to kiss Archie		
Archie gave Lela consent to kiss		

C Molly is in love! Molly is known to fall in love easily and desperately wants to get into a relationship. Molly sends multiple messages a day to Tayler and usually gets a one-word reply. Molly takes this as a positive and tells everyone they are together.

	Yes	No
Molly gained consent to tell everyone		
Tayler gave consent to tell everyone		

WORKSHEET 5.11 CONTINUED

D Alex is at a party and has had a few drinks. Alex gets up the confidence to see if Sam wants to go outside with them and get cosy. Sam leads Alex out and starts kissing Sam.

	Yes	No
Sam gained consent to kiss Alex	☐	☐
Alex gave consent to be kissed	☐	☐

2 Do all these scenarios warrant getting consent? Why/Why not?

____

____

____

____

3 What do you need to do to gain consent?

____

____

____

____

4 Where could you go if you needed help in these situations?

____

____

____

____

WORKSHEET 5.12 POWER

5

SB Pages 214–5

1 Provide three examples of how people use their power for good.

2 List three examples of how people abuse their power.

3 Provide three examples of when power is needed (e.g. by police to stop a criminal).

4 **Recall** examples of power that exist in your day-to-day life at school (e.g. teachers enforcing correct uniform).

Recall to remember; present remembered ideas, facts or experiences

AC

5 **Reflect** on whether you see power exist out in the streets or community places (e.g. shopping centres, movie theatres).

Reflect on to think about deeply and carefully

AC

6 **Justify** whether you have power. Where are you able to exert your power?

Justify to give reasons or evidence to support an answer, response or conclusion

AC

7 Provide examples of young people in the media exerting their power. What impact does this have on adults, do they listen, and do they do anything about it?

WORKSHEET 5.12 CONTINUED

8 Draw a mind map with all of the relevant people in your life (for example, family, friends, teachers, your coach and anyone else you come in contact with on a regular basis). Put yourself in the middle of the page and list/draw these people around the outside of the map.

a Take a blue pen and draw a line between you and each person that you have equal power with.

b Take a red pen and draw a line from you to each person that has power over you.

c Take a black pen and draw a line from you to each person that you have power over. Some people might have more than one line drawn to them.

Consider to think deliberately or carefully about something, typically before making a decision

9 **Consider** why some people might want to exert power over you.

____

____

____

____

____

Propose to put forward (e.g. a point of view, idea, argument, suggestion) for consideration or action

10 **Propose** what you would do if a person is exerting power over you and you believe they are wrong about something.

____

____

____

____

____

9780170465533

WORKSHEET 5.13 RESPECTFUL RELATIONSHIPS

5

SB Pages 210–17

1 Define a respectful relationship.

2 Sometimes people think there are different expectations for males and females in a relationship. Identify some of these gender stereotypes that males and females are 'supposed' to do in relationships.

3 **Discuss** how gender stereotypes influence a respectful relationship.

Discuss to talk or write about a topic, including a range of arguments, factors or hypotheses

AC

4 Propose how we can influence people not to use these stereotypes in relationships.

5 List three signs you might see if a friend was in an abusive friendship/relationship.

9780170465533

WORKSHEET 5.13 CONTINUED

6 Consider why people stay in an abusive friendship/relationship.

7 List three ways you could help a friend in an abusive friendship/relationship.

8 Discuss the following: It is better to have one or two good friends than 50 friends.

WORKSHEET 5.14 CONFLICT RESOLUTION HELP

Pages 222–3

1 Using Figure 5.32 on page 222 as a reference, fill in the bubbles below with reasons that there may be conflict in a friend or a partner relationship. For each conflict identify a person and/or a place that you could go to if you, or someone you know, needed help with that conflict. For example: your friend getting a new partner and not spending as much time with you might be a conflict. You could talk to your friend about this, and possibly go to other friends to spend some time with them. Or you might talk to a trusted adult for advice.

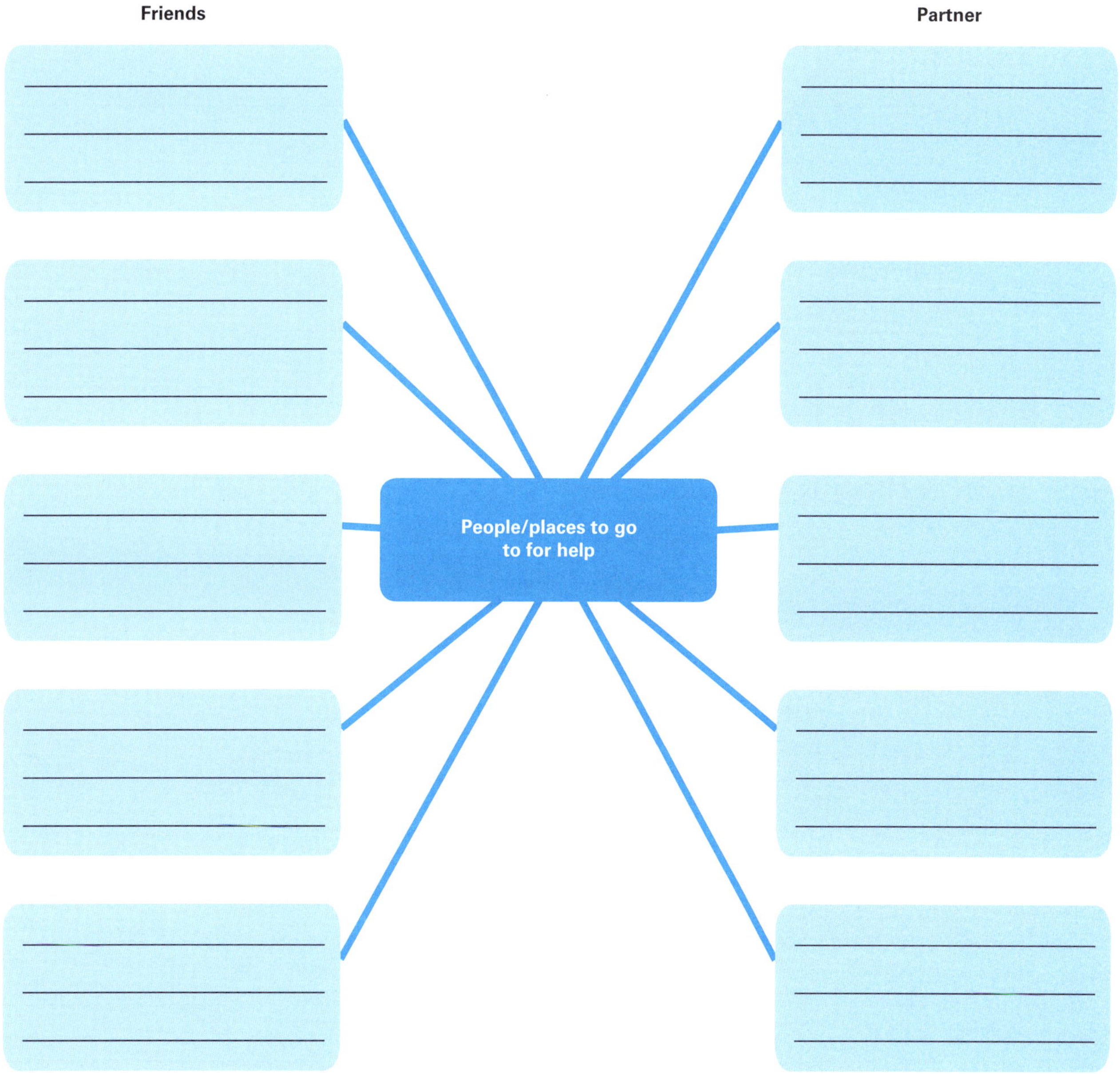

Friends

Partner

People/places to go to for help

CHAPTER 5 REVIEW

Reflect upon your learning in this chapter. Completing the following sentences will help to structure your thoughts about puberty and positive relationships.

1 Something I learnt about puberty is:

2 Something I learnt about respectful relationships is:

3 Something I learnt about power and privilege is:

4 One thing that surprised me was:

5 One thing I still wonder about is:

6 One thing I now want to know is:

9780170465533

6

THINK SAFE, ACT SAFE, BE SAFE

WORKSHEET 6.1 BULLYING

Page 228

Bullying is a clear and unacceptable abuse of power. Complete the following fishbone diagram to demonstrate you have considered some of the factors associated with bullying.

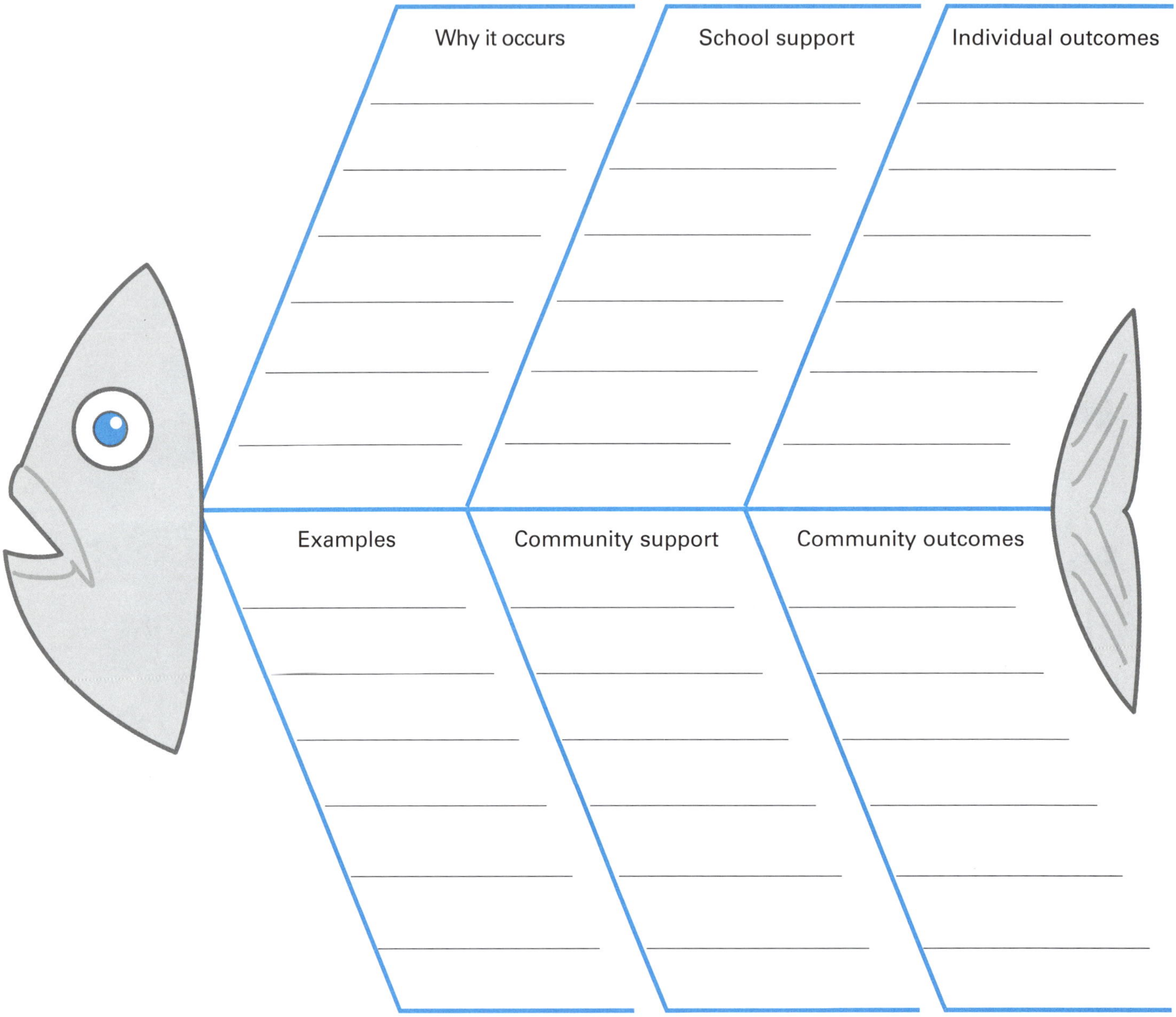

Why it occurs

School support

Individual outcomes

Examples

Community support

Community outcomes

9780170465533

WORKSHEET 6.2 CYBERBULLYING

SB Pages 230–3

Your task is to prepare an argument for a class debate on the topic 'Cyberbullying is worse than face-to-face bullying'.

In the table below, list three points that the 'affirmative' side might present to show why they agree with the statement. On the opposite side, list three points that the 'negative' side might present to argue why they believe that face-to-face bullying is worse than cyberbullying.

Cyberbullying is worse than face-to-face bullying	Face-to-face bullying is worse than cyberbullying
1	1
2	2
3	3

WORKSHEET 6.3 CYBER SAFETY

Page 233

A KWFL table (what you already **K**now, what you **W**ant to know, how you will **F**ind out and what you have **L**earnt) is a fun way of taking stock of 'where you are at' in certain areas. Complete the first three columns of the KWFL table. Then conduct the necessary research to complete the last column. Present your information via a poster, tri-fold information brochure or web page.

Consideration	What you already **K**now	What you **W**ant to know	How you will **F**ind out	What you have **L**earnt
Basic safety tips to protect yourself online				
Appropriate and ethical use of ICT and online communication				
Maintaining private information and your 'digital footprint'				
Defining cyberbullying and where a victim can get help				
Protecting your computer from viruses and setting up a firewall				

9780170465533

WORKSHEET 6.4 SHARING INAPPROPRIATELY

Page 238

1 Everything we share online leaves what we refer to as our 'digital footprint'. Your digital footprint — especially what you say and do on social media — may shape what people think of you, both now and in the future. That's why it's important to consider what kind of trail you are leaving, and what the impact might be.

In the footprint below, list types of personal details you have seen online, or heard people have shared online, that you think are either inappropriate or private. List the type of things you would not want others to share online about you.

2 An inappropriate photo or private information that is shared online with friends could end up damaging your chances of getting:

a a position of leadership within your school (SRC rep, house/sport captain, etc.)

b a part-time job

c a place on a sporting team

d a job interview

e entry to an overseas country.

discuss to talk or write about a topic, taking into account different issues or ideas

Discuss the potential effects of each of the above possibilities.

a ______________________________

____

____

____

b ______________________________

____

____

____

c ______________________________

____

____

____

d ______________________________

____

____

____

e ______________________________

____

____

____

9780170465533

WORKSHEET 6.5 PERSONAL SAFETY

6

SB
Pages 240–3

There are three steps that will help minimise your chance of harm in risky situations.

1. THINK

Think about any potential for a situation to become unsafe or risky. Listen to what your body is telling you: faster heart rate, increased sweating or an 'uneasy feeling'.

2. EVALUATE

Assess the level of risk associated with a situation. Quickly decide whether the situation is risky, but under control, or whether it is too risky and potentially harmful for you and your friends.

3. ACT

Act quickly to remove yourself from situations where you have determined your safety is at risk and things could go wrong for you or your friends.

Complete the following table, which considers different environments where your personal safety might be challenged:

- In the 'Best case scenario' column, provide two coping and helpful strategies.
- In the 'Worst case scenario' column, provide potential negative outcomes.

Scenario	Best case scenario	Worst case scenario
1 Your picture shared on Instagram		
2 Travelling to the city on public transport by yourself		
3 Wagging school (truancy)		
4 Catching up with another group of teens at a nearby river/beach		

Investigation skills

WORKSHEET 6.6 DRINKING TO HYDRATE

SB Page 248

Hydration is important, especially when you are involved in physical activity. Survey your class to find out how many students follow the recommended strategy in the table below.

1 Count the number of responses for each hydration strategy and convert this into a percentage by using the formula below:

$$\frac{\text{Number of students who use the strategy}}{\text{Number of students in the class}} \times 100 = \text{\% of students in the class who use the strategy}$$

Complete the following grid before you make your calculations.

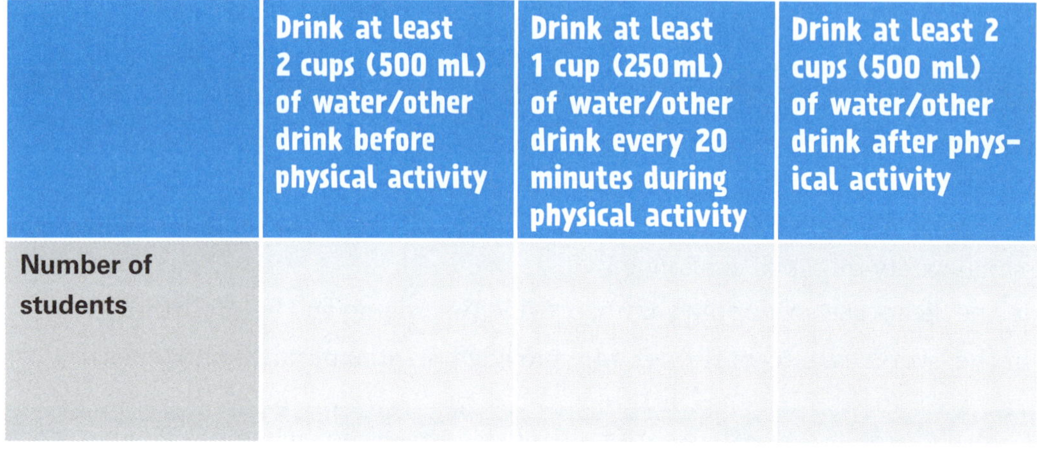

	Drink at least 2 cups (500 mL) of water/other drink before physical activity	Drink at least 1 cup (250 mL) of water/other drink every 20 minutes during physical activity	Drink at least 2 cups (500 mL) of water/other drink after physical activity
Number of students			

2 Using your calculations, graph your results, as three columns, below.

9780170465533

3 **Summarise** your findings and the recommendations you would make to your class about improved hydration strategies if you were a sports scientist.

summarise to give a brief statement of a general theme or major point/s

AC

4 How are 'sports drinks' such as Powerade or Gatorade different from 'energy drinks' such as Red Bull and V? You may need to discuss with others in your class, or do some research.

5 Are sports drinks worth the extra money, compared with simply going to the tap and drinking water?

WORKSHEET 6.7 ULTRAVIOLET RADIATION

Page 250

Australia has high rates of ultraviolet radiation (UVR) and associated skin cancers.

1 In the space below, create a storyboard for a TV 'infomercial' about the dangers of UVR and how people can protect themselves while outdoors. In each of the boxes headed 'Scene', note the message that you want to get across to the audience. In the boxes headed 'Words/Information', write any lines you want to be spoken or included in the 'infomercial'.

Infomercial title:

Scene 1	Scene 2	Scene 3	Scene 4
Words/Information	**Words/Information**	**Words/Information**	**Words/Information**
Scene 5	**Scene 6**	**Scene 7**	**Scene 8**
Words/Information	**Words/Information**	**Words/Information**	**Words/Information**

9780170465533

2 Quite often advertisers use 'shock tactics' to highlight the dangers associated with certain behaviours. Examples of these include the images on cigarette packets and the TV commercials that show negative outcomes for people who text or speed while driving or who have a poor diet and do not exercise.

a Do you believe these shock tactics are changing people's behaviours and improving their health? Discuss.

b Imagine you have been asked to design a shock tactics campaign to change people's complacent behaviour around the dangers of UVR and exposure to the sun. Briefly outline three different 'shocks' you would include in your campaign.

Shock tactic 1

Shock tactic 2

Shock tactic 3

WORKSHEET 6.8 WATER SAFETY

Pages 251–4

While Australia is an island with lots of beaches, both open and protected, there are also many lakes, rivers and other opportunities for people to have fun in and around water.

Unfortunately, water-related drownings occur every year.

Your task is to design a tri-fold pamphlet that focuses on water safety at one of the following three environments:

- swimming pool
- beach
- lake/river.

In the space below write down your ideas as if this was a 'brainstorming' or planning session. The first three columns represent the front of the pamphlet and the second three columns represent the back of the pamphlet – don't worry too much about how they fold on each other. Once you have finished your plan/sketch/information, swap with a classmate, who will then critically, but respectfully, offer you some suggestions to make improvements. You will do the same for them. If you agree that the suggestions are worth taking up, make the changes and then go away and create your tri-fold pamphlet. If you do not believe the suggestions will lead to improvement, respectfully argue your point of view.

When creating your pamphlet, make sure you acknowledge your sources of information (if relevant).

Front

Back

WORKSHEET 6.9 SPORT SHOES

Page 255

When you walk into a sports store, you are likely to see hundreds of different types of sport shoes. Sometimes, people choose their shoes because of the way they look, rather than how they perform or whether they are suited to a specific type of sport. You might be wearing a pair of shoes that look good, but are not suited to the sport you are playing or the surface you are playing on.

1 Conduct a survey in your class, perhaps at the end of a Physical Education class. Find out how many students are wearing sport shoes that are suitable for the activity they have just done. Count the number of responses for each shoe type or feature and convert this into a percentage by using the formula below:

$$\frac{\text{Number of shoes that display the feature}}{\text{Number of students in the class}} \times 100 = \begin{array}{l}\text{\% of students in the class whose}\\ \text{shoes display the feature}\end{array}$$

Complete the following grid before you make your calculations.

	Shoes have adequate cushioning/shock absorption in the heel	Shoes have adequate ankle support due to lacing design	Shoes are relatively lightweight	Shoes have adequate grip/tread design
Number of students				

2 Graph your calculations below. You did a similar exercise in Worksheet 6.6 so this time you have only been provided with the x- and y-axes of your graph. You will need to provide a heading and label both axes.

WORKSHEET 6.9 CONTINUED

3 Briefly discuss why AFL players need to have three or four different pairs of football boots at the beginning of each match. What determines which pairs they wear, and why might they change these during the match?

discuss to talk or write about a topic, taking into account different issues or ideas

AC

4 **Discuss** the following statement: Cross trainers are better than sport-specific shoes because they are so adaptable.

5 It is trendy to have laces tucked into the side of runners, rather than lacing them up. Discuss potential problems this may cause someone playing basketball or netball.

9780170465533

WORKSHEET 6.10 FIRST AID

It is likely you will need to apply basic first aid at some point in your life. Examine the illustration below and identify all the dangers that you can see. Justify the order of care you would provide if you were the first person to arrive on the scene (who you would attend to first, then next, etc.). Record your response in the table on the next page.

Pages 257–60

Dangers	I would attend to groups of people in the following order (1 = most urgent; 4 = least urgent)	My reason for treating the casualties in the order chosen

WORKSHEET 6.11 SPORT INJURIES

Pages 260–1

Physical Education classes provide students with positive learning experiences through active play, competition and involvement in physical activity. However, it's likely that on any given day, a student will be injured while participating in activities such as:

- basketball/netball
- softball/baseball
- football/soccer
- athletics/gymnastics
- hockey/lacrosse.

Choose one of the physical activity pairs above and identify a potential injury that might occur in each. Each activity must have a different type of potential injury. For each different injury, clearly state the steps you would take to provide or assist with first aid. Use the table below to organise your answers.

Activity selection	Potential injury and how it is likely to occur	First-aid steps
Basketball/netball		
Softball/baseball		
Football/soccer		
Athletics/gymnastics		
Hockey/lacrosse		

CHAPTER 6 REVIEW

Self-reflection at the end of units of work is a very useful way to gauge what you now know, what you enjoyed, what you need to find out and so on.

Complete the following statements relating to the work covered in this chapter on thinking, acting and being safe.

1 I learnt:

2 I really didn't understand:

3 I found these things interesting:

4 I enjoyed working with:

5 The next group that this is presented to should:

6 I would like to find out more about:

7 My favourite activity was:

8 It may have been good to also consider:

9780170465533

7

THE GREAT OUTDOORS

WORKSHEET 7.1 SLIDING DOORS ROLE-PLAYS

This activity highlights the importance of adequate preparation before you venture outdoors.

In this activity, groups of three or four students are to create a role-play based on a situation that could arise in the outdoors – e.g., the consequences of going out for a hike without proper planning or preparation. You will need to identify the beginning, the middle and the end of the scenario, and then perform either the beginning, the middle or the end of the scenario to the class.

Each section of the role-play might last for 30 seconds to a minute. The rest of the class must determine what might have happened in the other sections or what could happen at the end. As a challenge to the class, another two groups are invited to role-play the missing two sections.

The 'sliding doors' part of the role-play comes when the opposite occurs – appropriate and adequate planning. The group then presents a role-play that demonstrates a very different outcome to the previously poorly planned one.

Timing of outing	Poorly planned outing	Appropriately planned outing 'sliding door'
Before		
During		
After		

9780170465533

WORKSHEET 7.2 THE MENTAL AND PHYSICAL BENEFITS OF BEING OUTDOORS

CASE STUDY

Shutterstock.com/Michaeljung

How spending time outdoors can improve your health and wellbeing and boost your productivity and creativity.

After weeks and in some cases months of being cooped up due to COVID restrictions, relegated to a daily commute from bed, to couch, to kitchen, one thing we could all use a little more of is fresh air. That's because as we've all stayed closer to home, our routines have become more and more limited – a trend that has ripple effects on both our physical and mental health.

But going outside has more to offer than just a change of scenery – increasingly, greater exposure to nature and the outdoors has been shown to be associated with better overall health and wellbeing. It boosts our immune systems, lowers blood pressure and – in a time defined by its uncertainty – could be just what we need right now.

Now, more than ever, it'll help us cope.

Having some contact with nature can reduce chronic stress, explains Dr. Candice La Lima, PhD, a clinical psychologist with Northwell Health. 'It reduces symptoms of depression, sadness, low self-esteem, anger, and can improve overall mood'.

Our bodies physically become calmer while in nature. Our heart rate, muscle tension and production of stress hormones reduce – meaning we're able to process our surroundings in a less agitated state, and in turn feel more at ease. This sense of calm as a result of experiencing nature is especially helpful during today's widespread sense of collective hardship.

'During COVID, people are dealing with isolation from others, disruption from their daily routines and uncertainty about the future', La Lima says. 'Each of which can certainly impact our mental well-being, causing depression, irritability and anxiety.'

By experiencing the novelty that taking a walk in nature has to offer, we're better able to blow off some of that steam. It helps us to create space from the negative emotions associated with being confined to our homes.

Your productivity and creativity may depend on it

Given that a number of us are now expected to work at home, maintaining focus and productivity are more crucial than ever.

'When we're inside all the time, it can reduce our sense of self-worth', La Lima says. 'This in turn leads to a lack of productivity and impacts our focus or capacity to keep our attention on one task for a long period of time. When we experience depression and anxiety, they serve as barriers against our ability to think outside of the box'.

Spending time in nature can act as a reset button of sorts, allowing our brains to focus in a new way and return to our tasks with a renewed sense of energy.

'Taking a few minutes to step away from our daily routines can help us to gain that clarity of thought, improve our concentration, mood – and all this is helpful for promoting optimal productivity and creativity', says La Lima.

It's all about mindfulness

To make our time spent in nature more beneficial, we have to practice developing that calm sense of focus.

'Society promotes multitasking, like doing work while listening to a podcast or watching a show and at the same time exchanging text messages', La Lima says, 'but we're actually more effective if we limit stimulation to one source at a time'.

Try taking a walk outside without your cell phone, or without your headphones plugged in. Take slow deep breaths, allow your eyes to wander, listen to the sounds around you. To keep your mind stimulated, embark on a new route each week. Discover the things you never once noticed right in your own neighborhood.

It's accessible to all

While not everyone has the ability or proximity to large, open swaths of nature to take walks outside, there are still ways to incorporate it into your everyday routine.

'For those that aren't able to physically connect with a large outdoor natural space, there is research that tells us that visual exposure to nature – even if we aren't physically engaged with it – has beneficial effects on our mood and well-being', La Lima says.

This means looking out the window and taking particular notice to what it is you see. Mindfully connecting to each detail, even in small pockets of nature like a house plant or indoor herb garden.

When we allow ourselves to embrace all nature has to offer, we quiet that internal chatter and make space for better physical and emotional well-being

From 'The mental and physical benefits of nature during COVID-19', by Aisha Beau Johnson, The Well by Northwell, 24 April 2020.

 9780170465533

7

WORKSHEET 7.2 CONTINUED

Read the article extract above, then use the following PMI (Plus, Minus and Interesting) table to explore it in more depth by identifying pluses, minuses and interesting things (or implications) associated with the topic.

Plus	Minus	Interesting/Implications

Investigation skills

WORKSHEET 7.3 TOURIST BROCHURE

SB Pages 277–80

1 Research recreational activities in your State; you could use the State Tourism Bureau as a starting point. Choose one type of outdoor recreation that you believe would be popular among overseas tourists.

2 Research the work of organisations that are involved with efforts to reduce pollution and improve waste management and environmental awareness in your State.

3 Create a brochure advertising the outdoor recreational activity that you chose in question 1. Ensure that the brochure highlights the need for minimal impact and sustainability, and specifically targets what international tourists can do to help. Organise your ideas and layout in the space below before you create the final product.

Front

Back

Investigation skills

7

WORKSHEET 7.4 OUTDOOR RECREATION SURVEY

Many recreational activities are possible in the great outdoors, and they have various benefits. Create your own outdoor recreation survey about how people use the outdoors and what they gain from it.

1 Use a website such as KwikSurveys, which is free after you register, to create your online survey. Here are some suggestions for questions to include on the survey.
   - What is your age and gender?
   - What outdoor activities have you participated in?
   - Of the activities that you have participated in, which did you enjoy the most and why?
   - On average, for how long did you participate in each activity?
   - What costs were involved with the most enjoyable activity?
   - Who encouraged you or what motivated you to participate in this activity?
   - Which of the following health benefits did you gain from your experiences: physical, social, emotional, spiritual or environmental?
   - What do you consider are the major barriers that prevent people from experiencing the outdoors more frequently?

2 Using the results of your survey, answer the following questions.
   a What was the most popular outdoor activity? Did this differ between genders?

   ____

   ____

   b Which activity had the longest duration?

   ____

   c What were the most expensive and least expensive activities?

   ____

Being involved in active outdoor recreation or adventure activities has many benefits, mainly for the individual, but for society as well. There are benefits in the following four areas.

Benefits for the individual

- Connection with other people and communities
- Enhanced capacity to nurture self
- Positive impacts on cardiovascular health, cancer prevention, diabetes and mental health
- Increased skills (physical, problem solving, self-awareness)

Community/social benefits

- Improved quality of life
- Healthier families
- Enhanced health outcomes
- Improved physical, mental, social, community and environmental health and wellbeing

Environmental benefits

- Increased individual connection with nature
- Enhanced awareness of the natural environment

compare
to observe or note how things are similar or different

3 **Compare** the results of your survey to the summary of benefits above.

identify
provide an answer from a number of possibilities

4 **Identify** three major barriers to participating in the outdoors. Propose solutions to overcome the three barriers.

9780170465533

WORKSHEET 7.5 OUTSIDE GEOCACHE SEARCH – A TREASURE HUNT OR IS IT REALLY GEOCACHING?

SB Pages 282–4

Remember when you played 'Marco Polo' at the pool? It is a game many young students play where someone has their eyes closed (or is blindfolded) and they need to tag someone saying 'Marco' and 'Polo' as the finder gets closer or further away.

You are going to play a different version of this game. Divide into groups of five. Two students wait outside the classroom (they will be the 'finders' or 'geocachers') while the teacher tells the rest of the team members they need to hide a 'cache' outside.

Some geocaching etiquette

- There is a need for stealth so that you don't give it away that there is a cache in that location.
- If more than one geocacher arrives at a site at the same time, the first person to see the cache stands aside to give the other person a chance to find it too.
- Consider the environmental stewardship element of geocaching, that is, take care not to damage the flora or fauna in any way.

Get geocaching

In groups of five, appoint two geocachers. Two geocachers go in search, with the three other class members following and giving directions. Ensure everyone knows where north, south, east and west are, because you will need to give/follow directions. Make sure your directions are as specific as possible, e.g. 'about ten metres east' or 'head south when you reach the corner'.

When your team finds the cache, sign the logbook, then re-hide it exactly where it was before, for the next cacher to find and sign.

Shutterstock.com/Lasse Hendriks

WORKSHEET 7.6 TRAIL RIDING

Trail riding is a popular recreational activity, especially with younger people. The following extract is part of a survey of community attitudes towards outdoor recreation along a horse-riding trail network in Queensland.

The activities surveyed were running, picnicking, mountain biking, four-wheel driving, trail riding, bushwalking, horse riding and dog walking.

- Most respondents tend to have a slightly positive or neutral attitude to most non-motorised activities.
- Motorised activities affected nearly all respondents.
- The few horse riders who used the park did not report conflicts with nearly all non-motorised activities, but were negatively affected by motorised activities.
- Activities considered as having the largest number of social and environmental impacts were trail bike riding and four-wheel driving. Commonly perceived impacts were making too much noise, potential for injury or collision, damaging plants or animals and frightening wildlife.

Source: Griffith School of Environment, 2013.

iStock.com/brians101

Alamy Stock Phooto/Konstantin Shishkin

Your teacher will tell you about De Bono's six thinking hats (looking at decisions from different perspectives). After the class is organised into groups, each person is then allocated a thinking hat. Your task is to read the information in the extract above and suggest ways in which some of the negative environmental and social impacts of the recreational activities might be reduced. Explore the information using the format below:

- the white hat person is to focus on the data available; they are neutral and objective
- the red hat person is to focus on intuition, gut reaction and emotion
- the black hat person is to focus on the negative aspects, difficulties, and weaknesses, providing logical reasons
- the yellow hat person is to focus on the positive, plus points, providing logical reasons
- the green hat person is to focus on being creative, providing ideas and alternatives
- the blue hat person is to focus on decisions about process – organising the thinking and planning for action.

WORKSHEET 7.7 LISTENING QUIZ

Page 289

Listening is an important aspect of communication but is often overlooked or neglected when people are trying to improve their communication.

1 Use the listening quiz below to determine how well you listen. Score yourself on the listening skill by awarding one point if you answer 'never', two points if you answer 'sometimes', three points if you answer 'often' and four points if you answer 'always'. Add up your scores to find out how well you listen.

Listening skill	Never (1)	Sometimes (2)	Often (3)	Always (4)
I try to give every person I speak to equal time to talk.				
I enjoy hearing what other people have to say.				
I wait until someone has finished talking so that I can then have my say.				
I listen when I do not particularly like the person talking.				
I listen even when I do not agree with what the person who is talking has to say.				
I stop what I am doing while someone is talking.				
I look directly at the person who is talking and give them my full attention.				
I encourage other people to talk by my non-verbal messages.				
I ask for clarification of words and ideas I do not understand.				
I respect every person's right to his or her opinion, even if I do not agree with them.				

Your score:

Key:

30–40: You are a great listener.

20–29: You are a good listener.

10–19: You are not listening well to others.

≤ 9: You are a poor listener.

WORKSHEET 7.8 VARIOUS WAYS TO COMMUNICATE INFORMATION TO OTHERS

Page 290

THE GOAL

The goal of the activity is to traverse, with eyes closed or blindfolded, a designated area full of obstacles without touching any obstacle or any person.

METHOD

1 Select a 'playing field.' Go outside, if possible, but this activity can be done inside, even in rooms with fixed furniture (which can become objects to be avoided).

2 Distribute 'mines', e.g. balls or other objects such as bowling pins, cones, foam noodles, chairs, hats, etc. – almost anything will work.

3 You need to lead a partner through a small minefield using verbal cues from the sidelines.

4 Partner 1 puts on blindfold and attempts to walk across the minefield safely without touching the 'mines.'

5 Partner 2 gives Partner 1 verbal cues to direct them safely across.

RULES

1 Start on one end and move across the area to the opposite end.

2 Count the points for each object touched on the way across.

3 If a major value object is touched, or another classmate is contacted the players must start over but keep the score of the first attempt to add to the second attempt.

EXTENSION/VARIATION: NON-VERBAL CUES

Objective: To lead the blindfolded teammate across the minefield using non-verbal cues (clapping, whistling, stomping, etc.).

How to Play: Teammates use their non-verbal communication skills to help their partner cross the mine field safely with the fewest number of mine/classmate touches.

The games can be played competitively, e.g. which pair is the quickest or has the fewest hits?

DISCUSSION

1 What was more effective, verbal or non-verbal communication? Briefly discuss why.

____

____

____

9780170465533

2 What effect did practice/repeat attempts have on performance?

This activity requires the application of key skills and, like any new skill, is quite difficult to begin with but becomes easier with practice.

3 What are some of the skills you have improved in as a result of more practice? These might be physical skills, such as playing a particular sport; musical skills from rehearsing with an instrument; writing skills; mathematical skills, etc.

Complete the graph below, which shows the relationship between amount of practice and performance levels.

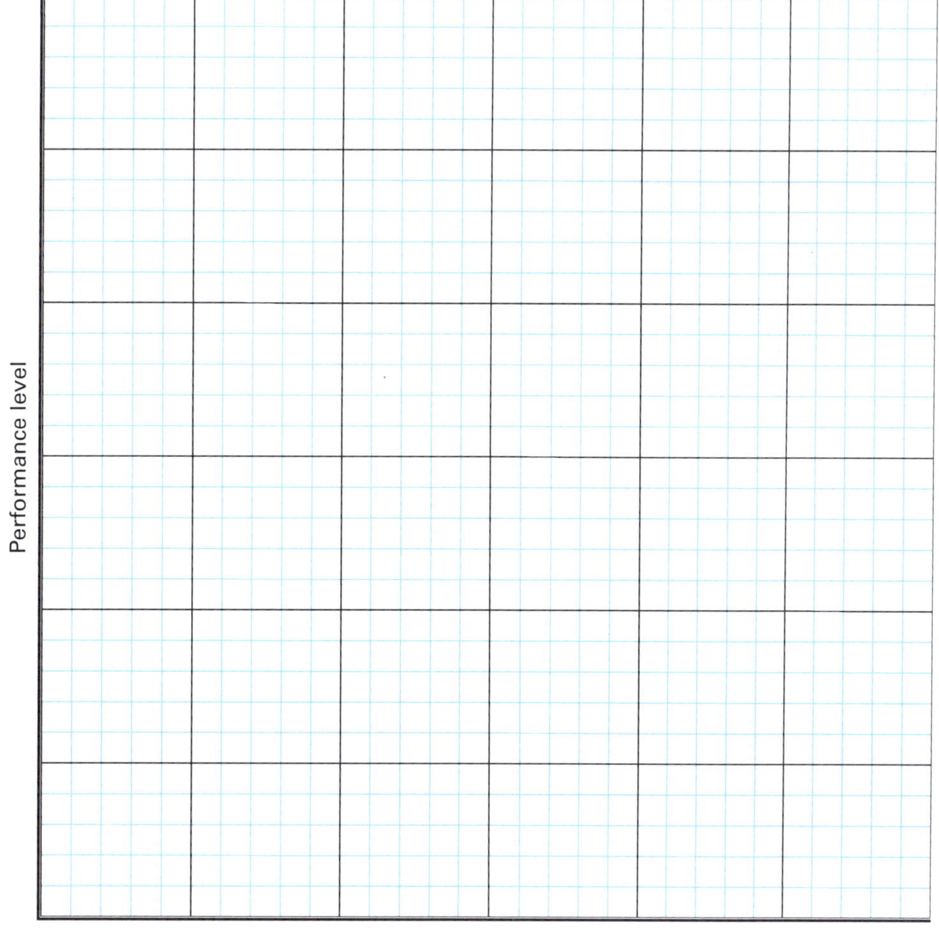

9780170465533

WORKSHEET 7.9 A SCORE COURSE IN ORIENTEERING

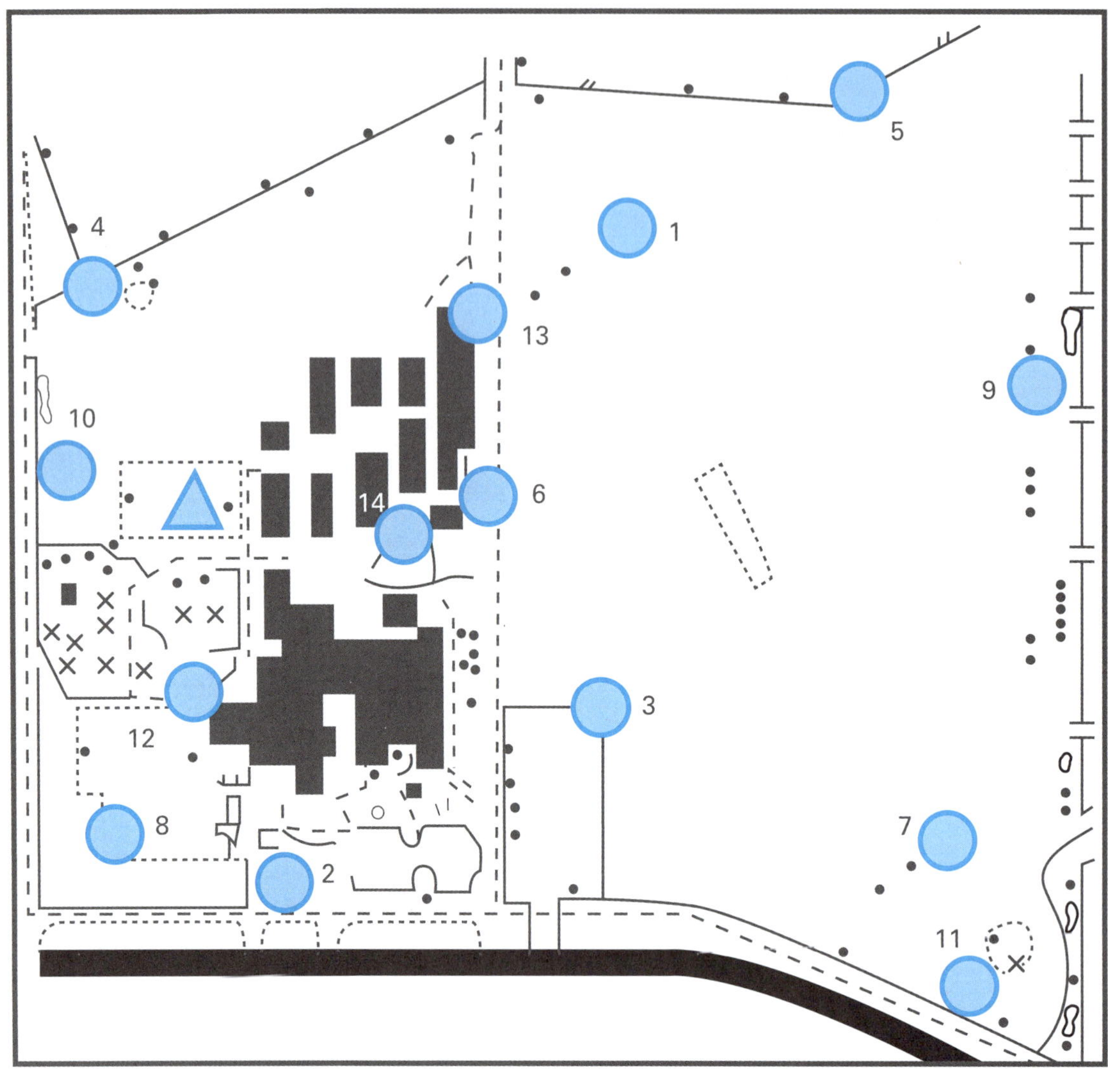

Pages 297–300

Your teacher will give you a map of your school, similar to the one above.

1 Under your teacher's supervision, set out 14–20 control points at different features around your school.

2 Use a table to allocate and record a points value to each control (e.g. 5/10/20 points). The aim is to gain the maximum score in the set time period (15 or 20 minutes, depending on the length and difficulty of the course).

3 All students start at the same time. There is no particular order you must follow; you just need to visit all the controls (numbers) in the shortest time.

4 A horn or siren will blow two minutes before your time is up, as a warning. You lose points for coming in late (e.g. 10 points per minute).

5 Good luck!

WORKSHEET 7.10 ORIENTEERING: THREE CLOVERLEAF COURSES

Your teacher will set up three courses, A, B and C, around your school. There are 5–8 controls per course. Each student will be given one page with three separate courses marked on it (a cloverleaf). The next three maps are just examples of how each course would work; do not use them, as they do not have your own school buildings and features.

SB
Pages 297–300

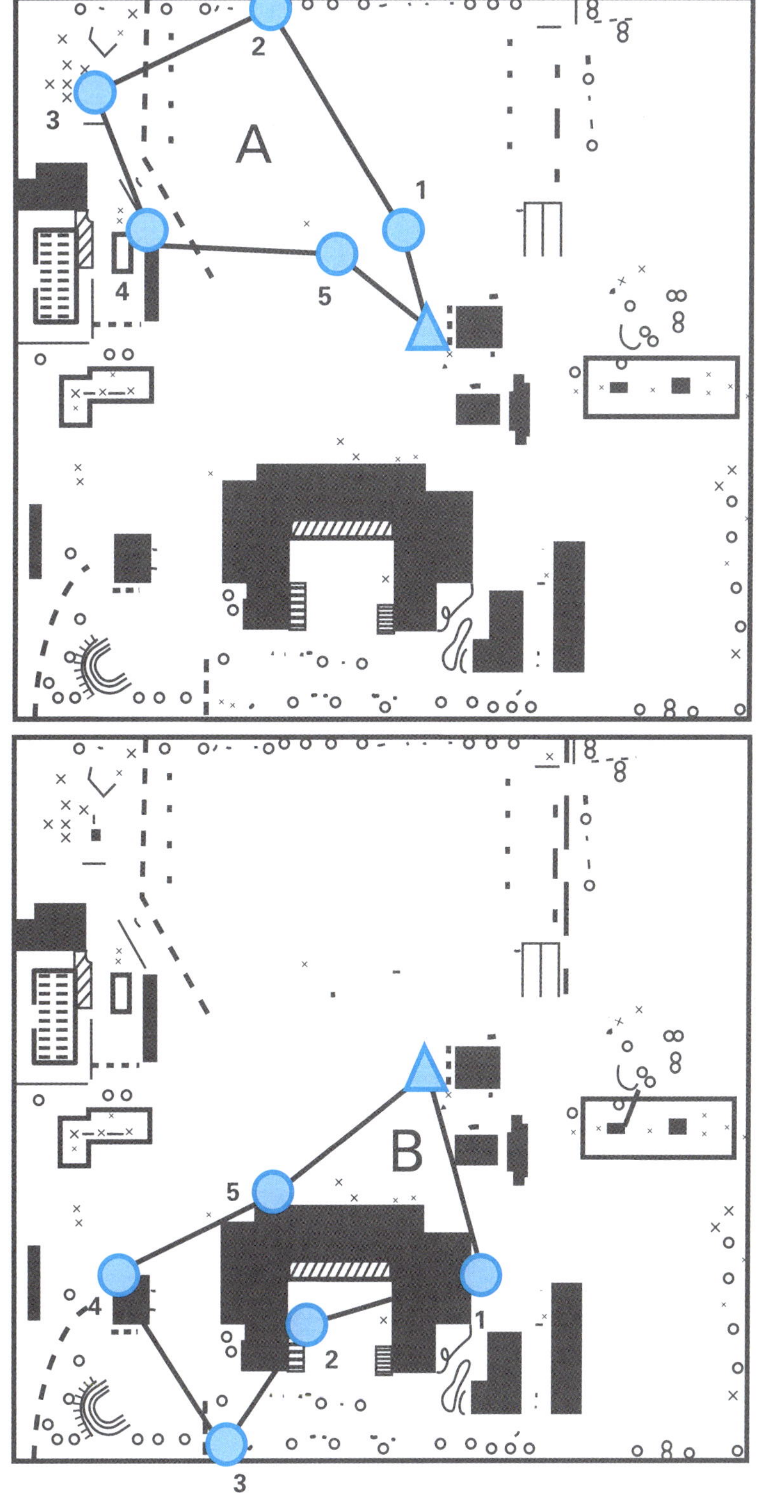

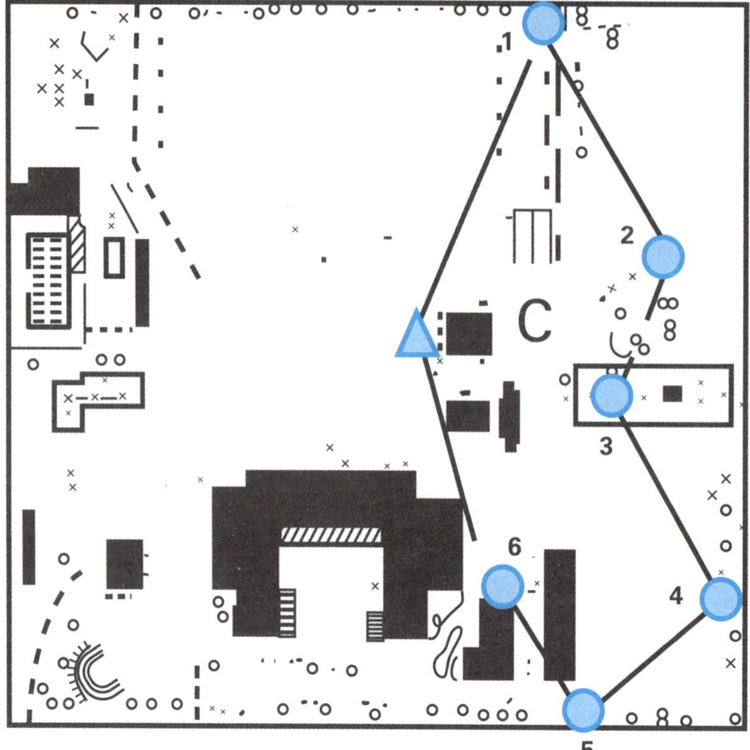

This activity can be done individually or in pairs.

1 For each course, start at the triangle in the centre of the map.

2 Three pairs of students start every five minutes, one pair per course.

3 You must visit the controls in the set order.

4 Once one course is completed, students move on to the second and third courses.

5 If you are fast, you will complete three courses in a session; if you are slow, you may do one or two.

WORKSHEET 7.11 HOW DID WE DO?

Page 301

Participate in a practical initiative game such as 'Madagascar'.

When you finish, make an analysis of the group's strengths, weaknesses, opportunities and threats (SWOT). Alternatively, you could assess your own performance. When all the analyses are completed, come together as a group and share your comments.

Weblink
Madagascar rescue game

SWOT analysis

Strengths	Weaknesses
Opportunities	**Threats**

Action plan

CHAPTER 7 REVIEW

Reflect upon your learning in this chapter. Completing the following sentences will help to structure your thoughts about outdoor physical activity.

1 One aspect of the chapter I found interesting was:

2 I learnt:

3 I now know:

4 I was surprised that:

5 I enjoyed:

6 I will always remember:

7 Two skills I have learnt are:

8 I still wonder:

9 I think we learnt about this topic because:

8

PLAYING THE GAME AND BEING A GOOD SPORT

WORKSHEET 8.1 PIERRE DE COUBERTIN

Page 311

Do a web search on Pierre de Coubertin. Use the following focus questions to summarise your understanding of his life and vision to restore Olympism across the globe.

1 Who was Pierre de Coubertin?

____

____

____

____

2 Why is the life and work of Pierre de Coubertin important to our understanding of sport and physical education participation?

____

____

____

____

Define to state meaning and identify or describe qualities

AC

3 **Define** the words 'elite', 'egalitarian' and 'truce'.

Elite: _______________

Egalitarian: _______________

Truce: _______________

4 How are the values of Olympism celebrated:

a in schools?

____

____

____

b in sport?

____

____

____

WORKSHEET 8.2 COMPETENT, LITERATE AND ENTHUSIASTIC SPORTSPEOPLE

Pages 312–15

One of the goals of Health and Physical Education in Australia is to provide you with the tools and information you need in order to become a competent, literate and enthusiastic sportsperson. Complete the following statements about your current level of competence, literacy and enthusiasm.

1 I am competent in the following games and sports:

____

____

____

2 I am literate in the following games and sports:

____

____

____

3 I am enthusastic about the following games and sports:

____

____

____

4 **Identify** people who you think are competent, literate and enthusiastic sportspeople.

Identify to recognise and name

AC

____

____

____

____

5 What evidence can you provide to support your conclusion that these people are competent, literate and enthusiastic in the sports you identified in the previous question?

____

____

____

____

Investigation skills

WORKSHEET 8.3

CREATING NEW GAMES: QUIDDITCH

1 Survey your class about their knowledge of the sport quidditch. Some questions you could ask are:

- Did you know quidditch was a real sport?
- Is quidditch a sport you are interested in playing?

2 Now ask your classmates whether they are interested in playing other sports, such as basketball, hockey or AFL.

Analyse to examine or consider something in order to explain and interpret it, for the purpose of finding meaning or relationships and identifying patterns, similarities and differences

3 **Analyse** the results of your question on sports your classmates are interested in playing by graphing the results below.

Page 317

Compare to display recognition of similarities and differences and recognise the significance of these similarities and differences

4 **Compare** the interest in quidditch to the interest in other sports. Suggest reasons for the results you find.

____

____

WORKSHEET 8.4 TARGET GAMES

Page 320

The aim of target games is to place a ball or other projectile near, in or on a target to achieve the best possible score. These games encourage and develop a high degree of precision in the skills of hand–eye coordination and concentration on a specific target.

All target games can be classified as either 'unopposed' target games or 'opposed' target games. Compare and contrast three different target games using a Venn diagram. Remember to say what is the *same* about each game (where the circles overlap) and what is *different* (where there is no overlap).

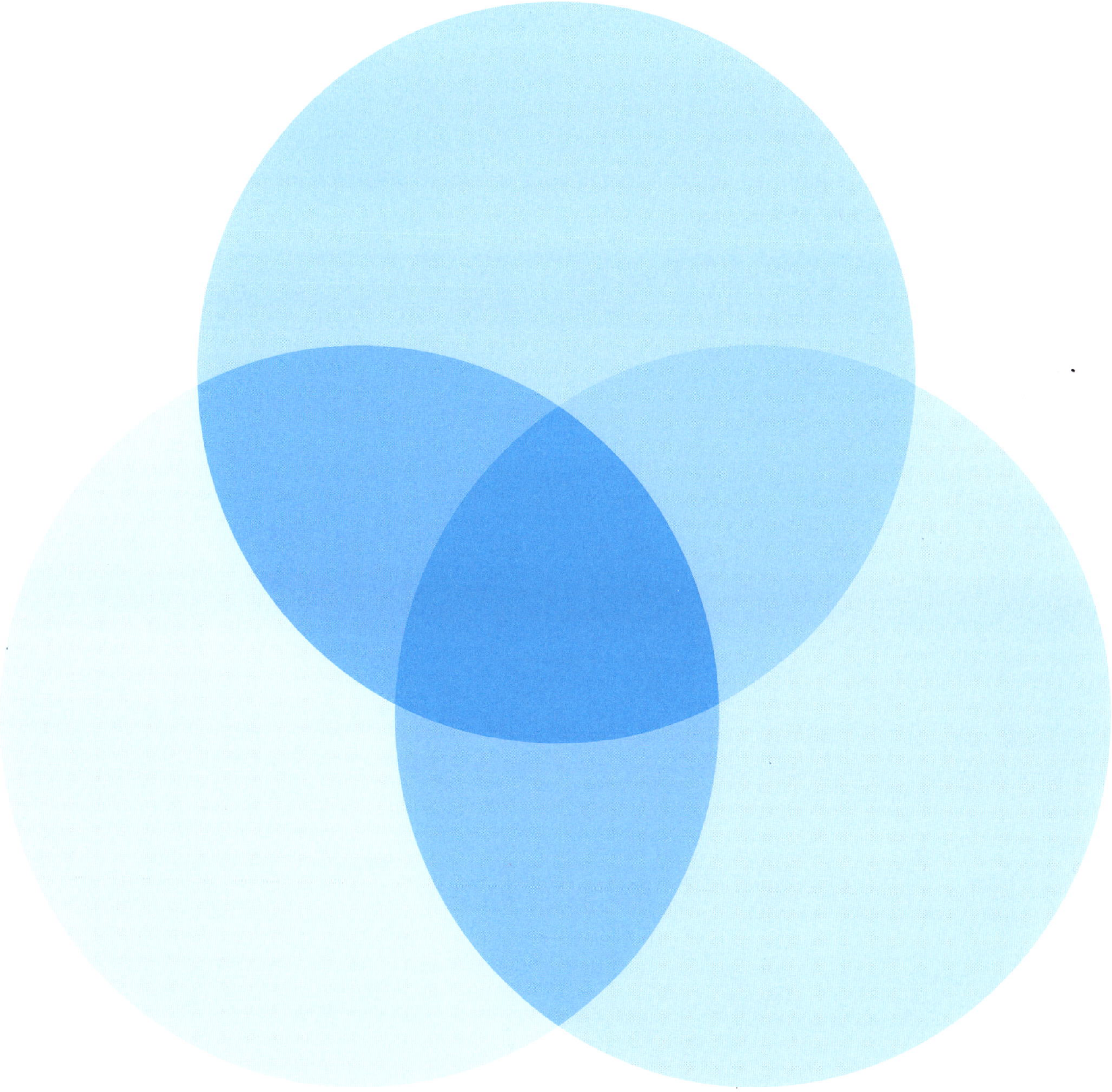

WORKSHEET 8.5 NET/WALL GAMES

Pages 322–3

The aim of net/wall games is to send a ball or projectile into an opponent's court so it cannot be played or returned by your opponent. This means open space on the opponent's court becomes your target. You need to be able to place a ball or projectile into the open space of your opponent's court while covering and defending as much of the open space as you can in your own court.

1 Use the KWFL chart below to assess and research your understanding of net/wall games. For each game listed, fill in each column of the table to show:
   - what you already **K**now
   - what you **W**ant to know
   - how you will **F**ind out
   - what you have **L**earnt.

2 Once you have planned and conducted your research, present your information in a poster, brochure or web page.

Game	What you already **K**now	What you **W**ant to know	How you will **F**ind out	What you have **L**earnt
Tennis				
Badminton				
Volleyball				
Sepak takraw				
Squash				
Racquetball				

9780170465533

WORKSHEET 8.6 STRIKING/FIELDING GAMES

8

Some of the key words used in striking/fielding games are listed in the first column of the table below. Fill out the second column with your own definition of each word, the third column with the *actual* definition and fourth column with a sentence that uses this word in the context of the sport in which it is played.

Pages 326–7

Word	What I think it means	What a reference says it means	How I can use it in a sentence about a specific striking/fielding sport
Innings			
Run			
Bunt			
Foul			
Strike			
Ball			
Delivery			
Block			
Over			
Boundary			

WORKSHEET 8.7 TERRITORIAL GAMES

Pages 327–8

1 Brainstorm the key decisions that a player needs to make in a 'territorial game' of your own choosing.

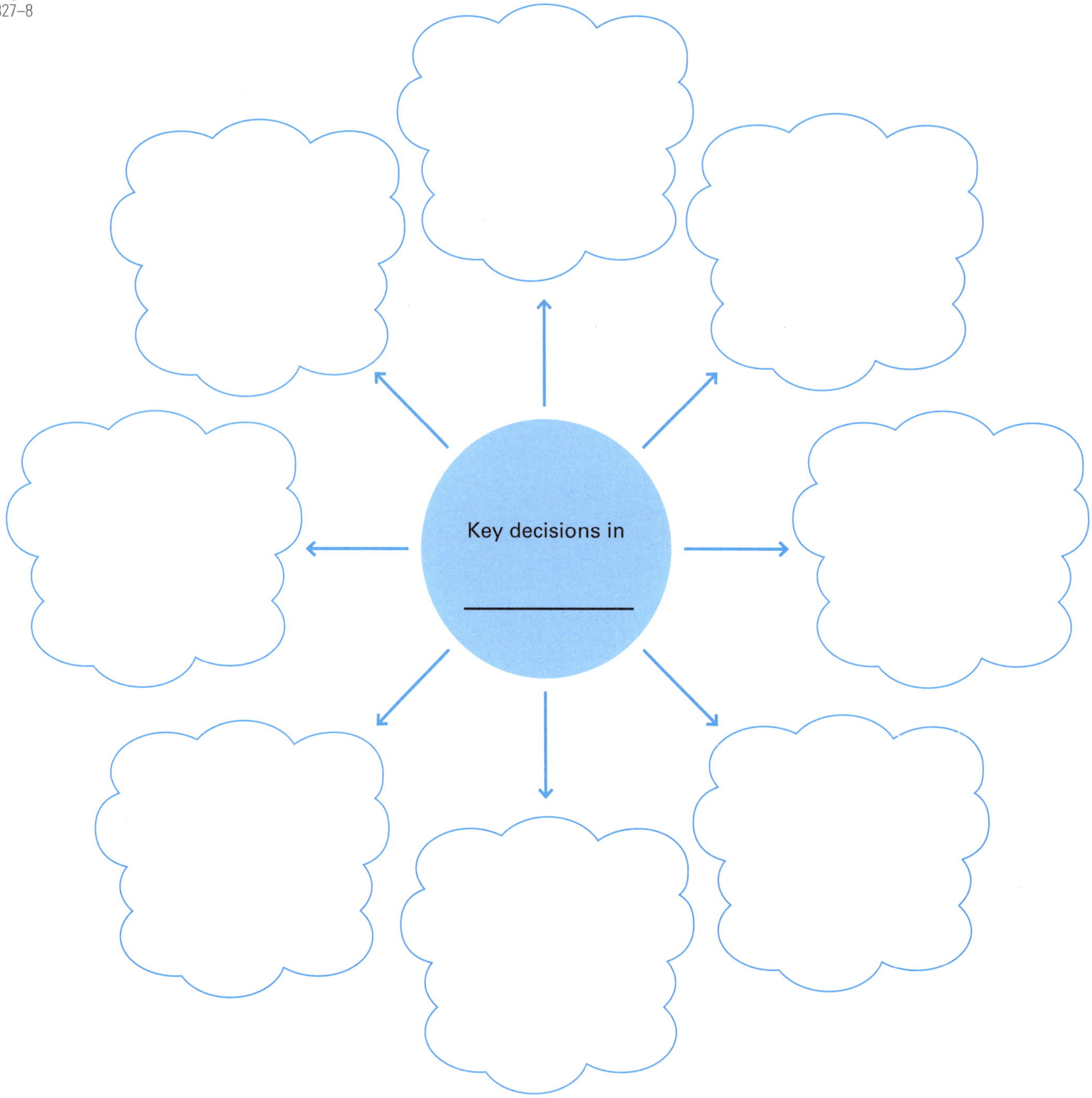

Key decisions in

____

WORKSHEET 8.8 DESIGN A SPORT SEASON

Use the table below to **explain** the pre-season considerations needed when designing a sporting season in your school.

SB

Page 328

explain to provide information that demonstrates understanding of reasoning and/or application

Task	Consideration	Action to happen at my school
Sport chosen		
Space		
Equipment		
Teams		
Duty teams		
Administration		

WORKSHEET 8.9 TEAMS, COMPETITION AND RECORDS

Pages 328–31

Using the information in your student book on team membership, competition and records, complete the following tasks.

Choose one activity from each row of the table below to complete.

	Challenging	More challenging	Most challenging
Row 1	Write a journal entry about when you needed to be reminded or prompted to participate. • What was the activity? • Why did you need to be reminded or prompted to participate?	Describe an instance when you took an active and leading role as a member of a sporting team. • What was the activity? • Why did you take an active and leading role? • How did it make you feel? • What did you learn from the experience?	Explain the social meaning you have personally drawn from being a member of a sporting team. • What were the positive and negative experiences? • How did these experiences make you a better person?
Row 2	Watch a televised sporting event and list the jobs of all the people you observe contributing to the success of that competition.	Think about all the roles and jobs that need to be fulfilled for a formal competition to occur in the sport of your choice. Summarise the roles and jobs, outlining the three most important aspects of each job.	Imagine you are given the task of running a school sporting carnival. How would you manage this?
Row 3	Read the sports section of a newspaper from your nearest city. Make a list of all the sporting records they contain. • Which sports are represented? • How many sports are only played by men/women?	Track the performance of your favourite sporting team over the last month from publicly available sources. Analyse their progress. • What can their success/downturn be attributed to, based on these records? • What additional information do you need to have to improve your analysis?	Create a tool (digital or non-digital) that helps individuals keep track of all the vital statistics and judgements that occur in a sport of your choosing.

9780170465533

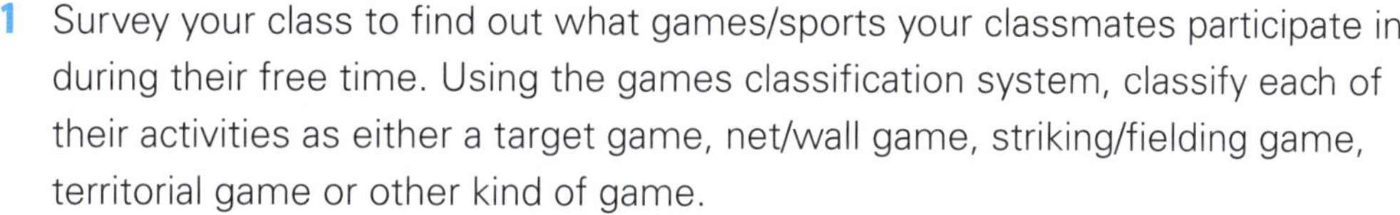

WORKSHEET 8.10 STUDENT SURVEY

SB Pages 320–8

1 Survey your class to find out what games/sports your classmates participate in during their free time. Using the games classification system, classify each of their activities as either a target game, net/wall game, striking/fielding game, territorial game or other kind of game.

2 Count the number of responses for each activity and convert the total number in each game to a percentage:

$$\frac{\text{Number of students who do the activity}}{\text{Number of students in the class}} \times 100 = \begin{array}{l}\text{\% of students in the class}\\ \text{who do the activity}\end{array}$$

3 Graph your results on the axes below. For further analysis, graph by a demographic suited to your class; for example, gender or cultural background. Construct a column graph, with one coloured column for each demographic per activity.

4 **Compare** how participation in physical activity changes depending on your chosen demographic.

____

____

____

____

5 Now survey the class to find out what roles students have adopted in different games/sports during their free time. Classify them into duty or team roles.

compare to display recognition of similarities and differences and recognise the significance of these similarities and differences

AC

DUTY ROLES

Most sporting competitions require a referee/umpire and a scorekeeper and may also include an assistant referee, touch judge, boundary umpire and judges. It is the duty team's responsibility to manage their participation during the competition.

TEAM ROLES

Team roles are non-player roles that encourage the functioning of individual teams. They may include coach, manager, trainer and fitness leader.

1 Count the number of responses for each type of role and convert the total number to a percentage:

$$\frac{\text{Number of students in the role}}{\text{Number of students in the class}} \times 100 = \begin{array}{l}\text{\% of students in the class}\\ \text{performing that type of role}\end{array}$$

9780170465533

2 Graph your results on the axes below. Construct a column graph, with one coloured column for each of the demographics you chose to analyse.

WORKSHEET 8.11 DESIGN YOUR OWN GAME

Page 342

Design a new game based on the games classification system.

1 Use the table below to explain each of the fundamental rules that exist in your game.

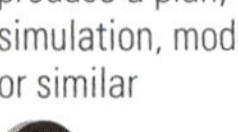

Design to produce a plan, simulation, model or similar

The four fundamental rules

1 *Scoring rules:* the skills needed to score points

2 *Player rights:* including equal chances to score points

3 *Freedom of action:* special actions players have with the ball that give the game its specific character

4 *Physical engagement:* ensuring the three rules above are respected through fair play and appropriate conduct

Scoring rules	Player rights	Freedom of action	Physical engagement

WORKSHEET 8.11 CONTINUED

2 In the space below, draw a diagram of the playing space of your new game. Ensure you include all the important line markings and a key that allows your teacher to interpret dimensions of the playing space, and the measurements of goals, targets, net heights and so on.

3 In your next Physical Education lesson, ask your teacher to pick the best games of your class and see if they can teach each game to the rest of the class using just the student's instructions. You may need to help them if they get it wrong, though! Good luck!

CHAPTER 8 REVIEW

Reflect upon your learning in this chapter. Completing the following sentences will help to structure your thoughts about playing the game and being a good sport.

1 I think we learnt about this topic because:

____

____

2 I wondered:

____

____

3 I learnt:

____

____

4 I now know:

____

____

5 I was surprised that:

____

____

6 I will always remember:

____

____

7 I still wonder:

____

____

8 I now plan to:

____

____

9

ENHANCING PERSONAL FITNESS THROUGH LIFELONG PHYSICAL ACTIVITY

WORKSHEET 9.1 LIFELONG PHYSICAL ACTIVITY

Pages 346–7

Lifelong physical activities are sometimes referred to as 'lifestyle' or 'lifetime' physical activities.

Lifelong physical activities:

- will improve your health if you perform them regularly
- should be performed daily
- can easily fit into your daily routines
- may feel light and easy for you while you are young, but will make you work harder when you get older
- are easy to perform even with a low personal level of fitness or skill
- use more energy than just sitting down
- should be kept up throughout your whole life
- require very little equipment
- can be performed anywhere, any time and even on holidays
- can be performed on your own or with one or two others
- that don't make you 'huff and puff' really hard are likely to become part of your daily routine.

1 Identify two activities that you currently participate in each week at home and school that could be classified as lifelong physical activities.

Home

____

____

School

____

____

2 In the box on the next page, draw or paste an image (photograph or diagram) of someone performing a lifelong physical activity. Label your image with five characteristics of lifelong physical activities from the list above. Refer to the example below, but use different characteristics from those provided in the example.

1 Expends more energy than rest
2 Can be performed at a moderate intensity
3 Can be performed on your own or with a small group of people
4 Can be performed with minimal equipment
5 Can be performed throughout the lifespan

9780170465533

3 Research techniques and cultural customs traditionally used by First Nations peoples to gather and prepare food. Discuss which of these activities would be considered lifestyle physical activities and why?

Investigation skills

WORKSHEET 9.2 LIFELONG PHYSICAL ACTIVITY SURVEY

The aim of this activity is to analyse what types of lifelong physical activities are participated in by your class.

1 Using the table below, survey the class to lifelong physical activities they participate in at least once per week.

Lifelong PA	Tally of participation	Class total
Badminton		
Aqua aerobics		
Bike riding		
Canoeing		
Dance		
Fitness (e.g. classes, aerobics, circuits)		
Cricket		
Household chores		
Gardening		
Golf		
Dog walking		
Flying disc		
Skateboarding		
Running		
Swimming		
Surfing		
Squash		
Tai Chi		
Tennis		
Walking		
Weight training		
Yoga		
Other: ________		
Other: ________		
Other: ________		
Other: ________		

2 As a class collate this data so you calculate a total for each lifelong physical activity. Use the table above to add the totals.

3 Graph the totals for each lifelong physical activity participated in by at least one person in your class. The *x*-axis should denote the proportion of students in your class that participate in a particular activity and the *y*-axis can display each lifelong physical activity. You don't need to graph activities where there was no participation (it must be done at least once per week on a regular basis).

4 Outline the five most common lifelong physical activities participated in by your class.

____

____

____

____

____

WORKSHEET 9.3 COMPONENTS OF FITNESS MIX AND MATCH

Pages 348–59

1 Examine the photographs in the table and find the best match for each component of fitness. Draw an arrow from the component term in the centre of the table to the most appropriate image representing this fitness component. Note: the sprinter is the only image that best depicts two components of fitness; the remaining images relate to one component each.

 Alamy/PCN Photography	 Alamy/blickwinkel	 Alamy/dbimages
 Getty Images/Jono Searle	Body composition Balance Reaction time Muscular power Speed Agility Coordination Aerobic capacity Muscular strength Muscular endurance Flexibility	 Shutterstock.com/trubavin

<table>
<tr>
<td>
Alamy/Action Plus Sports Images</td>
<td>Body composition
Balance
Reaction time
Muscular power
Speed
Agility
Coordination
Aerobic capacity
Muscular strength
Muscular endurance
Flexibility</td>
<td>
Shutterstock.com/Robert Crum</td>
</tr>
<tr>
<td>
Getty Images/SolStock</td>
<td>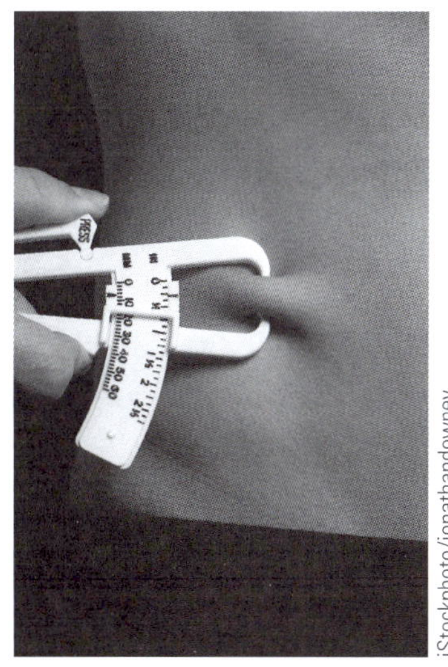
iStockphoto/jonathandowney</td>
<td>
Getty Images/GABRIEL BOUYS</td>
</tr>
</table>

2 If you had to find other images that show the following components of fitness, what images could you use? Describe an example for each.

Speed: ______________________________

Coordination: ______________________________

Body composition: ______________________________

Flexibility: ______________________________

Balance: ______________________________

WORKSHEET 9.4 DESIGNING A PHYSICAL ACTIVITY CIRCUIT AT HOME

Page 349

The aim of this activity is to design a design, apply and evaluate a physical activity circuit you could use at home, making use of recycled equipment or inexpensive sporting or household items.

Search around your home, yard and garage for any of the following items:

- brooms
- broom handles
- buckets and containers
- small balls (e.g. tennis balls)
- frisbees
- PET milk or cordial containers that could be filled with water
- hay bales, milk crates, cardboard boxes or other items that could be used as low steps
- pool noodles
- tarps, towels
- chalk
- ropes (e.g. skipping ropes).

1 In the space below design the layout of your circuit. Show where each station will be located and draw the equipment to be used. Describe briefly the activity to be performed at each station and label the equipment to be used, keeping in mind safety considerations.

Adapt your previously acquired knowledge and skills from circuits you have completed at school or in the community to your home circuit. Also use your imagination to create completely new ideas.

9780170465533

2 Describe where in your home or garden you could complete your physical activity circuit.

____

____

3 Use the blank table on the next page to summarise the activities in your circuit. Use a star or other graphic device to identify the fitness components that can be developed by the activities carried out at each station. Use the table below as an example.

2 mins per station × 2 circuit laps	Health-related fitness component developed				
Activity description	**Aerobic capacity**	**Muscular strength**	**Muscular endurance**	**Flexibility**	**Body composition**
Step-ups onto a wooden step	★		★		
Lifting two buckets filled with water using a bent-over rowing action		★			
Wrapping a towel around your feet and pulling your trunk towards your legs				★	
Skipping rope	★			★	★
Bicep curls using 2 × 2-litre PET containers filled with water			★		

WORKSHEET 9.4 CONTINUED

Complete this table for your physical activity circuit.

2 mins per station × 2 circuit laps	Health-related fitness component developed				
Activity description	Aerobic capacity	Muscular strength	Muscular endurance	Flexibility	Body composition

4 a Reflect on the design of your circuit, consider factors such as the selection of physical activities, the sequence (e.g. did the order allow you adequate rest for particular muscle groups?), what fitness components were addressed and the equipment selection. Discuss the strengths and limitations of your design and implementation of your circuit.

b Review your discussion and make a judgment about whether or not your physical activity circuit used an adequate range of different fitness components, was enjoyable and sustainable. (E.g. Would you consider doing it again regularly? Explain why or why not.)

5 Using the simple and inexpensive items listed on page 202, create a game you could play at home, either by yourself or with one or two other people. You need to create:

a a name for your game

b a description of the game (also use a diagram if possible)

c a list of equipment to be used

d the rules of the game

e a scoring system if relevant.

WORKSHEET 9.5 FITNESS CARTOONS

9

SB
Page 349

The aim of this activity is to locate and describe a cartoon that portrays a fitness component.

1 Search online for 'fitness components' or search online images for 'fitness cartoons'. Find a cartoon that depicts a particular fitness component. Save or print a copy of the cartoon. Complete the following, referring to the example on the next page if necessary.

   a Identify the relevant component of fitness depicted by the cartoon and classify it as a health-related or skill-related component of fitness.

   b Explain how the cartoon depicts the particular fitness component.

   c Describe one exercise you could use to develop this component of fitness.

   d Display the cartoons in your classroom or somewhere else in your school such as homeroom, gymnasium, locker bay, etc. Make sure you check with teachers first to obtain approval to 'post/display' these.

WORKSHEET 9.6 COMPARING MUSCULAR STRENGTH AND MUSCULAR ENDURANCE

Pages 349–50

The aim of this activity is to understand the difference between muscular strength and muscular endurance.

Participate in activities that require the use of muscular strength and muscular endurance (refer to the table below) either during class time or for homework.

Muscular strength	Muscular endurance
Pushing against a wall using maximum force.	Doing continuous sit-ups for 1 minute.
Pushing against a partner safely.	Doing continuous push-ups for 1 minute.
Lifting a heavy weight using correct technique (bent knees, keep back straight, and tense your stomach and core muscles).	Doing continuous squats or lunges for 1 minute.
Picking up several shopping bags and put them into a car.	Using a rowing machine or swimming laps.
	Using an exercise bike/cycling.
	Walking up several flights of stairs.

1 Define muscular strength and muscular endurance and then check your answers with your teacher. Correct them if they were wrong.

Muscular strength

____

____

____

____

____

____

Muscular endurance

____

____

____

____

____

____

9780170465533

WORKSHEET 9.6 CONTINUED

9

2 Use the information in the table below to explain how having good muscular strength and muscular endurance can help you to perform everyday tasks.

Having good muscular strength ...	Having good muscular endurance ...
Allows you to move objects or body parts with force (useful in sport and everyday life).	Allows you to exert a force repeatedly, like running up several flights of stairs or mopping the floor.
Helps in weight loss and maintaining a healthy body weight, due to the increase in your metabolism.	Allows you to repeat movements for a long period without too much fatigue.
Allows you to control your own body weight when lifting or bracing against a force such as a tackle in AFL or rugby.	Assists you to avoid injury.
Is useful in sports such as weightlifting and tackling, in resistance training and for lowering something heavy or pushing.	Helps in activities such as rowing, cycling, swimming, push-ups and sit-ups.

WORKSHEET 9.7 HOW FLEXIBLE ARE YOU?

Page 351

The aim of this activity is to measure your hip and trunk flexibility.

Complete the following static flexibility test for hips and trunk. All you need to conduct this test is a sit-and-reach box and a ruler. The table below displays norms in centimetres for the modified sit-and-reach test.

Males (cm)	Rating	Females (cm)
≤ 5	Poor	≤ 5
–5–1	Fair	6–9
2–5	Average	10–12
6–9	Good	13–16
> 9	Excellent	> 16

Method

- Complete a 10-minute warm-up and remove your shoes.
- Sit with your legs fully extended and knees locked.
- Place the soles of your feet against the sit-and-reach box.

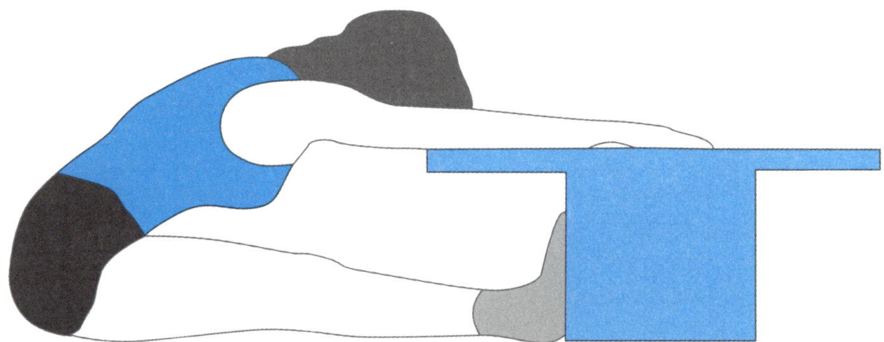

- Place your hands one on top of the other and lean forward.
- Reach forward as far as possible, with your fingertips sliding along the ruler on the box.
- Before you measure the final number in centimetres you must hold your final position for 2 seconds.
- Perform the test three times and calculate the average.

$$\frac{\left(\underset{\text{Score 1}}{\_\_\_\_\_}\right) + \left(\underset{\text{Score 2}}{\_\_\_\_\_}\right) + \left(\underset{\text{Score 3}}{\_\_\_\_\_}\right)}{} = \frac{\text{Total score}}{3} = \text{Average score}$$

1 Compare your calculated average to the norms in the table above and state your rating.

____

2 Explain why it is important to be flexible.

3 Discuss whether it is possible or not to be flexible in one muscle group but not another.

4 Go online and research the 'shoulder rotation test'. Describe how this test is used to determine flexibility and outline the procedure.

5 Describe why shoulder flexibility would be important during the following movements:

a Putting a shirt on

b Bowling in cricket

WORKSHEET 9.8 MEASURING YOUR BODY'S RESPONSE TO PHYSICAL ACTIVITY

Pages 359–62

The aim of this activity is to practise measuring your heart rate and evaluate which physical activity is associated with the highest intensity.

MEASURING YOUR HEART RATE (PULSE)

- **Radial pulse:** place two fingers on your wrist, just below the base of the thumb.
- **Carotid pulse:** place two fingers (index and middle fingers) on either side of your neck. Do not press hard. If you measure your carotid pulse on the right side, use your right hand and press gently.

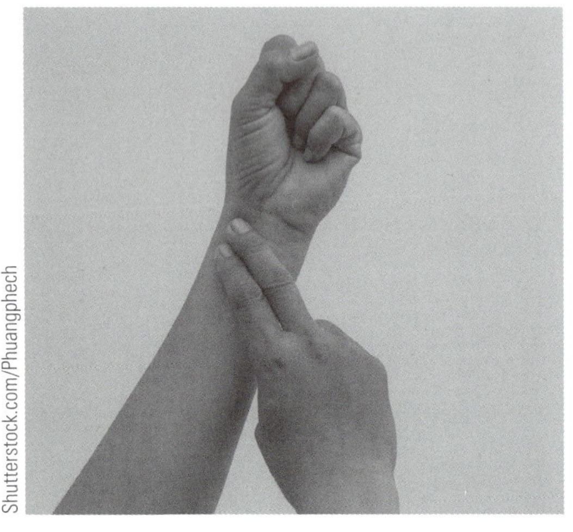

Shutterstock.com/Phuangphech

Shutterstock.com/New Africa

1 Practise taking your carotid pulse at rest.

2 Complete several of the following physical activities and immediately afterwards use the table to record your carotid pulse in beats per minute (bpm). You will need to perform each activity, pause and take your pulse and record it in the table below.

Your pulse (bpm)	Your activity
	Resting, sitting and relaxing
	Walking slowly for 100 metres
	Jogging slowly for 200 metres
	Sprinting for 100 metres
	Shooting hoops for 2 to 5 minutes
	Shadow boxing for 1 minute
	Completing 10 full or modified (on knees) pushups slowly with correct technique
	Skipping with a rope for 1 minute

9780170465533

WORKSHEET 9.8 CONTINUED

9

3 Graph your heart rate for each activity either by hand or electronically. An example graph has been provided.

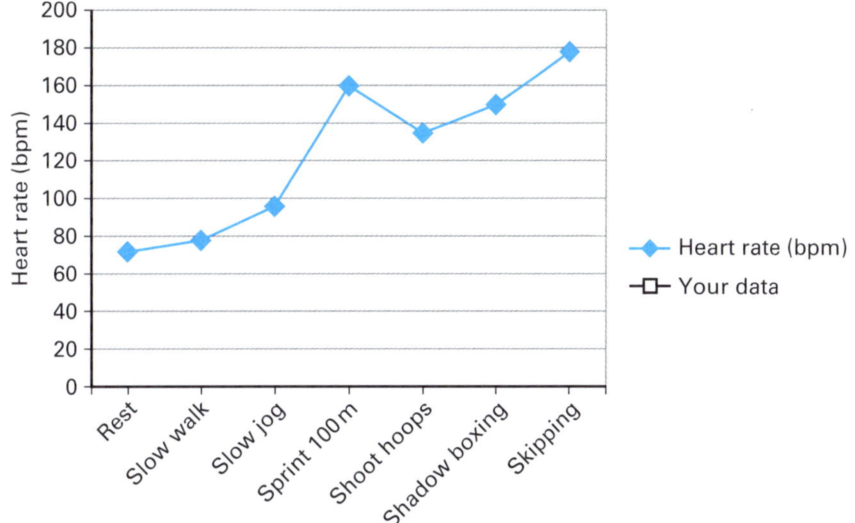

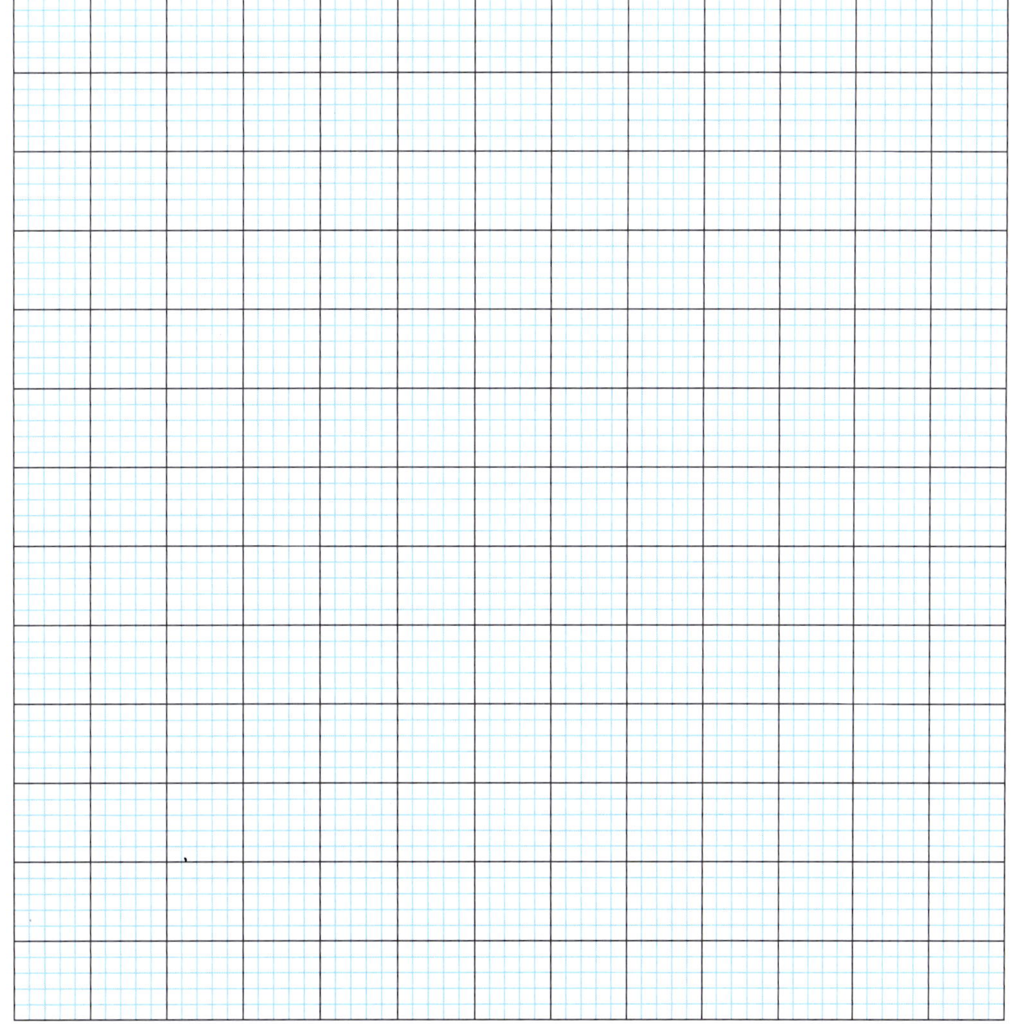

WORKSHEET 9.8 CONTINUED

4 Determine which three physical activities associated with the highest heart rate (bpm).

5 State how much your heart rate increased in total from resting levels to the most vigorous physical activity you performed (e.g. during skipping my heart rate increased by 49 beats per minute).

6 Describe the strengths and limitations of measuring your body's response to physical activity by measuring your carotid pulse.

7 Describe another method of measuring your heart rate other than measuring your carotid pulse.

WORKSHEET 9.9 TALK TEST

Page 361

The aim of this activity is to determine whether you are exercising at a moderate intensity by using the talk test.

Shutterstock.com/Rido

Participate in one of the following physical activities, gradually increasing the intensity:

- walking, marching on the spot
- doing step-ups onto a low bench
- walking up stairs
- jogging
- bike riding or using an exercise bike
- skipping with a rope
- dancing.

During the physical activity, have a conversation with your partner about any topic (e.g. the weather, what you did on the weekend, your favourite TV show or movie). Every few minutes, gradually increase the intensity you are working at until you can no longer keep talking.

1 State which physical activity you performed.

____

2 Describe how you felt when you were unable to hold a conversation (e.g. puffing hard).

____

____

____

____

WORKSHEET 9.10 CLASS QUESTIONNAIRE

Complete the CLASS Questionnaire individually thinking about a typical week for this time of year.

During a typical week what activities do you usually do?	Do you usually do this activity?	MONDAY–FRIDAY		SATURDAY–SUNDAY	
		How many times **Monday–Friday?**	Total hours/ minutes **Monday–Friday**	How many times **Saturday and Sunday?**	Total hours/ minutes **Saturday and Sunday**
E.g. Bike riding	No_1 Yes_2 (circled)	2	40 min	1	15 min
Aerobics	No_1 Yes_2				
Dance	No_1 Yes_2				
Calisthenics/gymnastics	No_1 Yes_2				
Tennis/bat tennis	No_1 Yes_2				
Aussie Rules Football	No_1 Yes_2				
Soccer	No_1 Yes_2				
Basketball	No_1 Yes_2				
Cricket	No_1 Yes_2				
Netball	No_1 Yes_2				
Baseball/softball	No_1 Yes_2				
Swimming laps	No_1 Yes_2				
Swimming for fun	No_1 Yes_2				
Down ball/four-square	No_1 Yes_2				
Tag/chasey	No_1 Yes_2				
Skipping	No_1 Yes_2				
Roller blading	No_1 Yes_2				
Scootering	No_1 Yes_2				

During a typical week what activities do you usually do?	Do you usually do this activity?	MONDAY–FRIDAY		SATURDAY–SUNDAY	
		How many times **Monday–Friday?**	Total hours/ minutes **Monday–Friday**	How many times **Saturday and Sunday?**	Total hours/ minutes **Saturday and Sunday**
Bike riding	No1 Yes2				
Household chores	No1 Yes2				
Playing on playground equipment	No1 Yes2				
Playing in a cubby house	No1 Yes2				
Trampolining	No1 Yes2				
Playing with pets	No1 Yes2				
Walking the dog	No1 Yes2				
Walking for exercise	No1 Yes2				
Jogging or running	No1 Yes2				
Physical education class	No Yes				
Sport class at school	No1 Yes2				
Walking to school (to and from school = 2 times)	No1 Yes2				
Cycling to school (to and from school = 2 times)	No1 Yes2				
Other (please state)	No1 Yes2				

During a typical week what other leisure activities do you usually do?	Do you usually do this activity?	Total hours/minutes Monday–Friday	Total hours/minutes Saturday and Sunday
E.g. TV/videos	No1 Yes2 (circled)	15 h	6 h 30 min
TV/videos	No1 Yes2		
Playstation/Nintendo/ computer games	No1 Yes2		

Telford, A, Salmon, J, Jolley, D and Crawford, D, 'Reliability and validity of physical activity questionnaires for children: the children's leisure activities study survey (CLASS)', *Pediatric Exercise Science*, vol. 16, no. 1, 2004, pp. 64–78

WORKSHEET 9.10 CONTINUED

discuss
to talk or write about a topic, taking into account different issues or ideas

1 Refer to the data in your completion of the CLASS measure, **discuss** whether you think you would meet the Australian 24-Hour Movement Guidelines for Children and Young People (5–17 years). Ensure you address the various components of the guidelines relevant to your age group, consider the frequency, intensity, duration and type of physical activities you engage in.

9780170465533

WORKSHEET 9.11 HOP TO IT

When creating a personal fitness/physical activity plan, there are two key cognitive strategies to be aware of:

- the risks of inactivity
- the benefits of being physically active.

There are many health benefits associated with regular physical activity. High-impact activities involve your body contacting either an object (such as hitting a punching bag when boxing) or the ground (when you are skipping with a rope). High-impact activities are excellent for developing bone strength. Examples of high-impact activities include:

- skipping
- jumping
- dancing
- landing from a height (below waist height)
- boxing or striking a ball with an implement
- hopping
- leaping.

1 Design and draw up a circuit/obstacle course you could do either at home or in the local park. It could be performed indoors or outdoors at home. Activities might include jumping over small homemade hurdles made out of broom handles on buckets. Include at least five high-impact activities in your obstacle course. You could get someone to film you completing the course or take photos of someone completing each obstacle. Make sure the obstacle course you design suits the space you have chosen and the equipment you have access to.

Here are six example activities.

Use the space below to design and create your obstacle course. Use a number key to show the order of rotation around the obstacle course. Describe each activity and any safety considerations.

WORKSHEET 9.12 CREATING A TV COMMERCIAL

9

Page 375

There are important cognitive and behavioural strategies you can use to develop a personal/physical activity plan. These include:

- **Cognitive strategies:** relating to understanding, comprehending and thinking about things (e.g. understanding the benefits of being regularly active or increasing awareness about the opportunities to be active)
- **Behavioural strategies:** relating to things we do (e.g. use a reward system).

1 Research a physical activity opportunity, in either your local community or school, that would be attractive to your fellow students. Create a television or radio commercial advertising this opportunity. The activity you select needs to appeal to someone your age.

Apply the following information when preparing the film your commercial:

a name of the physical activity

b location

c health benefits of participating in the physical activity

d when you can access the physical activity or facility

e cost and any other important information.

apply to use knowledge and understanding in response to a given situation or circumstance

2 You need to feature in your commercial at some stage. The commercial only needs to be a few minutes long. Ensure you use energising, motivational music and other special effects when editing your commercial. Your teacher may even select several commercials to screen in front of your class.

Shutterstock.com/Andrey_Kuzmin

Investigation skills

WORKSHEET 9.13 GETTING ACTIVE

One of the most powerful devices you can use to help increase your awareness of how physical activity you are each day is simply to count your daily steps.

1 Use a pedometer (or a smart watch that has a pedometer function) to record your daily steps for a week.

Monday: ______________________ total steps

Tuesday: ______________________ total steps

Wednesday: ____________________ total steps

Thursday: _____________________ total steps

Friday: _______________________ total steps

Saturday: _____________________ total steps

Sunday: _______________________ total steps

Weekly total: __________________

2 Graph your step count for each day.

9780170465533

3 How many days did you exceed the 10 000 steps recommendation?

____

4 Discuss whether your step counts were higher on weekdays or weekends and **justify** why you think this is the case. Also comment if that week was a typical week for you.

____

____

____

____

____

____

justify to show how an argument or conclusion is right or reasonable

5 Collate the daily totals for your class and obtain an average for each day.

The formula is:

$$\frac{\text{Total steps of all students}}{\text{Number of students}} = \text{Average}$$

Write the averages below.

Monday: ______________

Tuesday: ______________

Wednesday: ______________

Thursday: ______________

Friday: ______________

Saturday: ______________

Sunday: ______________

6 Graph the average step count each day for your class.

7 Discuss whether the step count was high for your class on weekdays or weekends and justify why the step count was higher on weekdays/weekends.

9780170465533

WORKSHEET 9.14 USING SOCIAL SUPPORT TO GET ACTIVE

One of the most important strategies you can use to help you stick to your personal physical activity plan is to enlist social support. This simply means asking someone to be active with you. It might be a family member, a friend or a neighbour.

1 List physical activities you would like to do with someone else in the next month. With a partner, go through your lists and compare physical activity ideas. Report back and share at least two activities with the rest of the class or with another pair.

a ______

b ______

c ______

d ______

e ______

f ______

g ______

h ______

i ______

j ______

k ______

2 Discuss why you think people are more likely to be active if they have someone to be active with. Think about social factors, increased safety and encouragement.

______

______

______

______

______

______

______

______

WORKSHEET 9.14 CONTINUED

3 In each box, write a fun physical activity and the name of a person you could participate in this activity with.

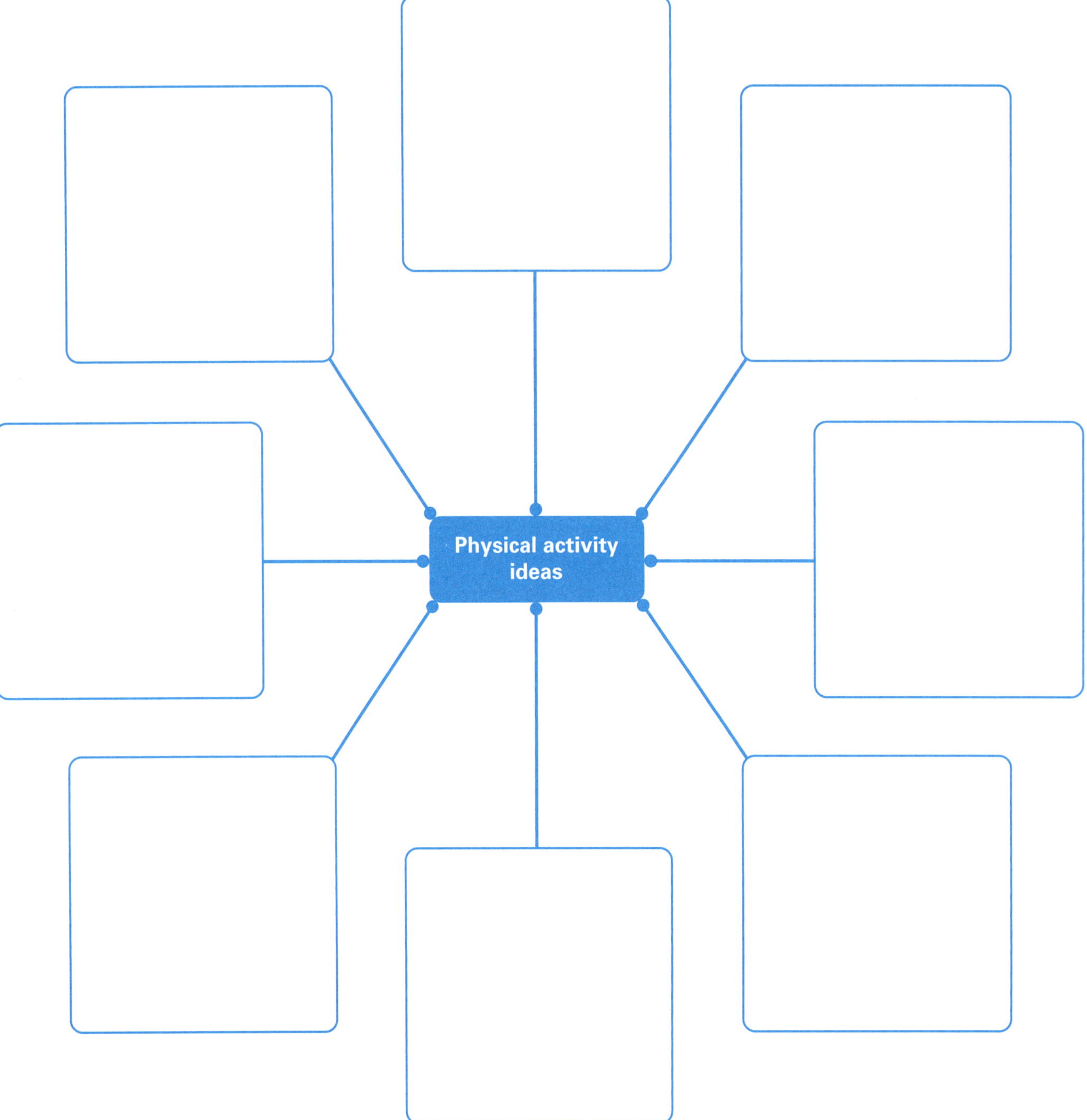

9780170465533

WORKSHEET 9.15 EVALUATING THE HEALTH OF YOUR SCHOOL

As a class, use this school health index to **evaluate** the 'health' of your school.

evaluate to examine and judge the merit, significance or value of something

Component	Excellent (3)	Adequate (2)	Needs improvement or under development (1)	Not in place (0)
Policies				
Policies relating to health are published and displayed around school				
Policies relating to nutrition are published. E.g. students know which foods can be brought from home				
Policies in relation to sun protection				
Policies about bullying prevention				
Student representatives have input into decisions that affect student health				
Students have access to physical activity (PA) facilities after hours				
Students have access to PA facilities during recess and lunch breaks				
Healthy food and beverages are sold in the school canteen				
Health and PE curriculum addresses nutrition, active lifestyles, wellbeing, and sex and drug education				
Amount of active homework				
School newsletters include ideas for being active outside school				
Active transport is encouraged				
Total score for policy				

WORKSHEET 9.15 CONTINUED

Component	Excellent (3)	Adequate (2)	Needs improvement or under development (1)	Not in place (0)
Social strategies				
Active role models during lunchtime, e.g. older students taking programs for younger students; teachers being active during yard duty				
Social dance opportunities				
Walking club/group				
Social sporting competitions at lunchtime				
Drama program				
Total score for social strategies				
Physical environment (built/man-made environment)				
Adequate PA facilities				
Easy access to drinking water				
Adequate bike racks				
Playground equipment				
Playground markings				
Walking trails nearby and in school grounds				
Physical environment (natural environment)				
Vegetable garden				
Accessible water features (e.g river, ocean, creek, lake or pond)				
Accessible hills or other bushland				
Accessible grassed areas				
Trees, logs and rocks in the school grounds				
Total score for built and natural environment				
Overall total				

Source: Associate Professor Amanda Telford, School of Education, RMIT University

9780170465533

WORKSHEET 9.15 CONTINUED

1 **Identify** three strengths in your school based on your audit.

identify to provide an answer from a number of possibilities; recognise and state a distinguishing factor or feature

AC

2 Discuss which aspects of the index you could use to improve your own health.

3 Identify three areas that are limitations/barriers to being active and in need of improvement in your school to support students to be active.

4 **Describe** a strategy for each area that needs improvement.

describe to give an account (written or spoken) of a situation, event, pattern or process, or of the characteristics or features of something

AC

WORKSHEET 9.16 FREQUENT MOVER PROGRAM

A powerful behavioural strategy for achieving success with your personal activity plan is to establish a rewards system. In this activity, think about the type of rewards you could use to encourage yourself.

Frequent flyer or loyalty programs reward customers with either points or credit towards either more flights or items such as vouchers, televisions, toasters, movie tickets, etc. The more flights taken or money spent on credit cards, the more rewards people accumulate.

The rewards system that you design should include the physical activities you need to do, or the sedentary behaviour you need to minimise/reduce, and the associated reward. For example, if you went for a walk around the park you might reward yourself with 10 points. When you have got to a certain number of points, you might use those points towards something of your choice such as watching your favourite movie. You may want to negotiate your reward system with a parent or guardian, especially if there is a cost involved. Certain devices (e.g. some smart watches) or apps enable you to record your activities which are then awarded points within rewards systems.

1 Complete the table below.

List five rewards that do not cost money (e.g. watching your favourite TV show, putting a smiley face sticker on a wall chart, adding points to your goal total)	List five physical activities that could be completed to obtain one of the free rewards (e.g. walking around the park for 20 minutes)
1 ____	1 ____
2 ____	2 ____
3 ____	3 ____
4 ____	4 ____
5 ____	5 ____
List five rewards that cost less than $5 (e.g. chocolate bar, smoothie, download of a new song, sticker)	**List five physical activities that could be completed to obtain one of the rewards under $5**
1 ____	1 ____
2 ____	2 ____
3 ____	3 ____
4 ____	4 ____
5 ____	5 ____

9

WORKSHEET 9.16 CONTINUED

List five rewards that cost between \$5 and \$25 (e.g. DVD, movie ticket, ticket for sporting event, sporting equipment)	List five physical activities that could be completed to obtain one of the rewards between \$5 and \$25
1 ______	1 ______
2 ______	2 ______
3 ______	3 ______
4 ______	4 ______
5 ______	5 ______

2 Create a continuum of rewards using a points system, such as the one below.

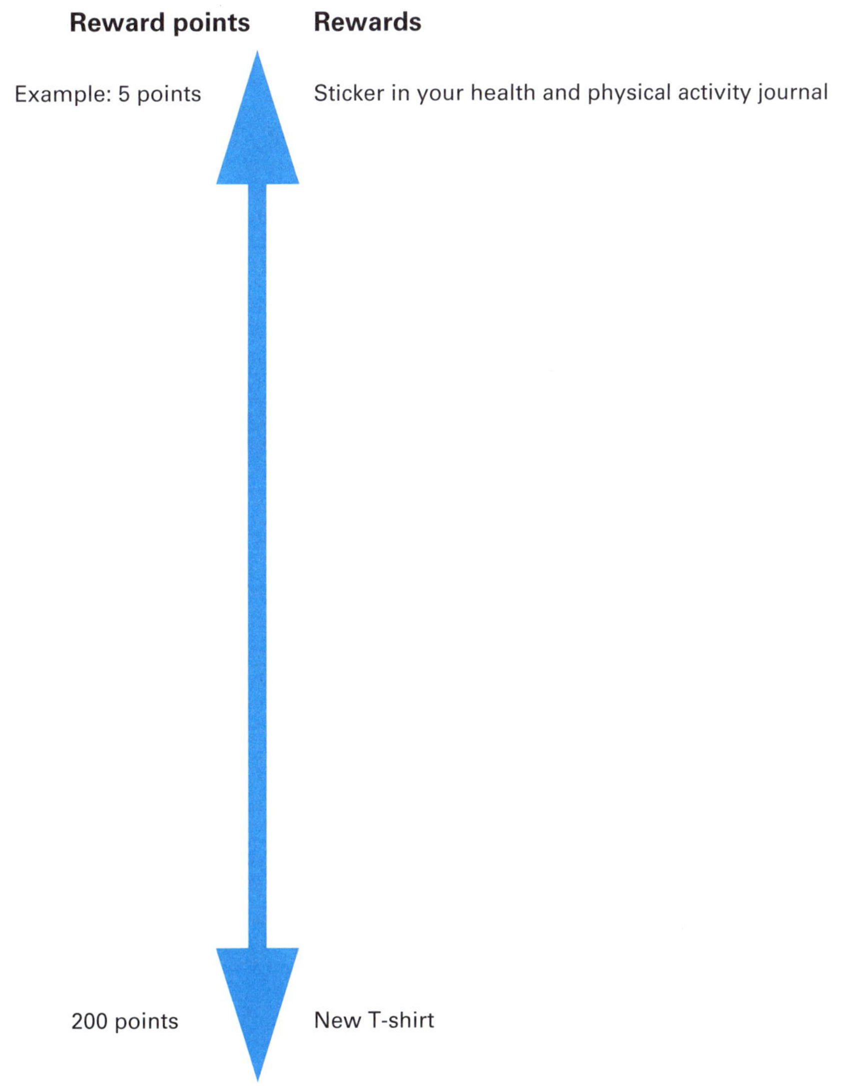

WORKSHEET 9.17 GOAL SETTING FOR SUCCESS

There are two very important behavioural strategies you need to use to succeed with your personal physical activity plan:

- making commitments
- goal setting.

We can use the acronym SMARTER to help us think about effective goal setting. Provide a description and example for each of the words below.

1 Goals need to be:

Specific

Measurable

Accepted

Realistic

Time phased

Exciting

Recorded

2 Outline one long-term goal that you could set and work towards to maintain a healthy and active lifestyle. For example, 'I am going to do something active for at least 60 minutes every day, for the next six months'.

3 Identify three short-term goals that you could work towards achieving in the next few days. For example, 'I am going to walk our dog for 20 minutes after school on Monday, Wednesday and Friday'.

a

b

c

4 Identify three short-term goals that you could work towards during the next three months. For example, 'I am going to walk to and from school one day a week for the next eight weeks of term'.

a

b

c

5 State a place where you could display your goals so you won't forget them, such as on the fridge or on a poster in your bedroom. Think about a place that would suit you best.

____

6 Describe how you will track whether or not you reach your goals.

____

____

____

____

7 Describe how you will remind yourself to be active and reach your goals. (Be specific; think about the reminder system you set up.)

____

____

____

____

It is important to develop realistic goals and display them somewhere you can see them. Are these examples of realistic goals?

WORKSHEET 9.18 GETTING ACTIVE AT HOME: ROLE-PLAY

For your personal physical activity plan to be a success, it is important to be able to identify the activity opportunities all around you.

In this group activity, imagine you have to self-isolate for a week at home due to the COVID-19 pandemic. You need to think about a simple physical activity you could do at home. Each group will be assigned an activity and you will need to come up with a fun role-play that outlines how to perform your physical activity at home.

Examples include:

- dancing during the commercial breaks on television
- doing step-up exercises while watching YouTube
- doing bicep curls (using a plastic 2-litre milk/cordial bottle) while talking on the phone (change hands halfway through)
- cleaning up the house quickly to make it a good workout
- vacuuming while listening to music
- doing push-ups and sit-ups between episodes of your favorite shows on TV
- chasing the dog or your little brother or sister around the backyard/house.

You could think about another activity you could do at home to get active that is not on this list and run the idea past your teacher.

1 Identify your favourite place at home to be active. Explain why.

____

____

2 Describe three physical activities you already do.

____

____

____

____

WORKSHEET 9.19 REMINDING YOURSELF TO BE ACTIVE

In this activity, you will think about a range of strategies you could use to remind yourself to be active. Using a reminder system is an essential part of a successful personal physical activity plan. You could even encourage some of your family members to use some of these strategies.

Things that can be used as reminders around the home appear in the table below.

1 Tick the strategies that you could use in your home.

Tick for yes	Example reminder strategy
	Leave your runners at the front door.
	Place a message on the kitchen fridge.
	Put a note on the noticeboard at home.
	Have a bag packed with runners and activity clothing in the hallway, ready to take to school or work.
	Put a note on the calendar in the family room or your bedroom.
	Put a poster on a pinboard in your bedroom.
	Ask a friend to text you or message you using social media.
	Set a reminder in your phone or using a physical activity tracker such as an app.
	Put a poster on the toilet door.
	Have your sporting equipment bag packed, ready to take to the park.
	Keep dog leads near the front or back door.
	Have exercise equipment accessible in your family room or lounge room; for example, do bicep curls while watching television.

2 Outline two more strategies you could use to remind yourself to be active.

WORKSHEET 9.19 CONTINUED

3 Draw your home and yard, and label six places where you could be active, such as the stairs, garden and so on. You could even use an app or Minecraft to illustrate a sketch of your home.

4 Add dot points for each place labelled which outline physical activities that could be completed within them.

5 Where in your home/yard are you generally the most and the least active?

6 Describe one activity you could implement in your home on your own and one you could participate in with at least one other person.

CHAPTER 9 REVIEW

1 Place a tick against each of the statements that apply to you.

Now I can:

- ◯ Identify and participate in lifelong physical activities that enhance my health-related and skill-related fitness and wellbeing.
- ◯ Outline health-related and skill-related fitness components and assess my flexibility.
- ◯ Create, apply and evaluate a personal physical activity plan and design fitness circuits at home.
- ◯ Use cognitive and behavioural self-management skills to maintain regular participation in physical activity, including goal setting, reminder systems, social support and reward systems.
- ◯ Evaluate health information and explore physical activity opportunities in my local natural and built environment; for example, within my school and local community.
- ◯ Measure my pulse.

2 Reflecting on my study about enhancing fitness through lifelong physical activity, I would like to learn more about the following:

9780170465533

10

JUST DANCE!

WORKSHEET 10.1 WHY DO PEOPLE DANCE?

Page 385

An important feature of human beings is the ability to communicate, not only by speech and writing, but by physical expression. While there are many ways for people to communicate, one of the most popular methods through the ages has been to dance.

The ability to speak allows people to communicate effectively, but can all your ideas, emotions and feelings be expressed through speech alone? The physical and visual portrayal of ideas, stories and emotions through movement can convey so much more than just words. For many people, dancing is about self-expression – how you feel usually influences how you move.

Since ancient times, dance has been used to convey how people think and feel. It can help to tell a story, to celebrate or to express culture, spirituality or emotion. Dancing can also provide many social opportunities.

Importantly for many people, dance is now becoming a very popular form of exercise. Dancing just to move your body can also benefit your health.

reflect on to think about deeply and carefully

AC

1 Working on your own, **reflect on** the following question: 'Why do you dance?'

____

____

____

____

examine to investigate, inquire or search into; consider or discuss an argument or concept in a way that uncovers the assumptions and interrelationships of the issue

AC

2 Now find a partner and **examine** the following questions together. Give three responses to each question.

a Why do people dance?

____

____

____

____

b Who or what influences your dance choices?

____

____

____

____

9780170465533

WORKSHEET 10.1 CONTINUED

10

3 As a class, explain your results and try to **classify** responses so you can create a class graph of the most frequent responses. Draw the graph in the space below.

classify to arrange, distribute or order in categories according to shared qualities or characteristics

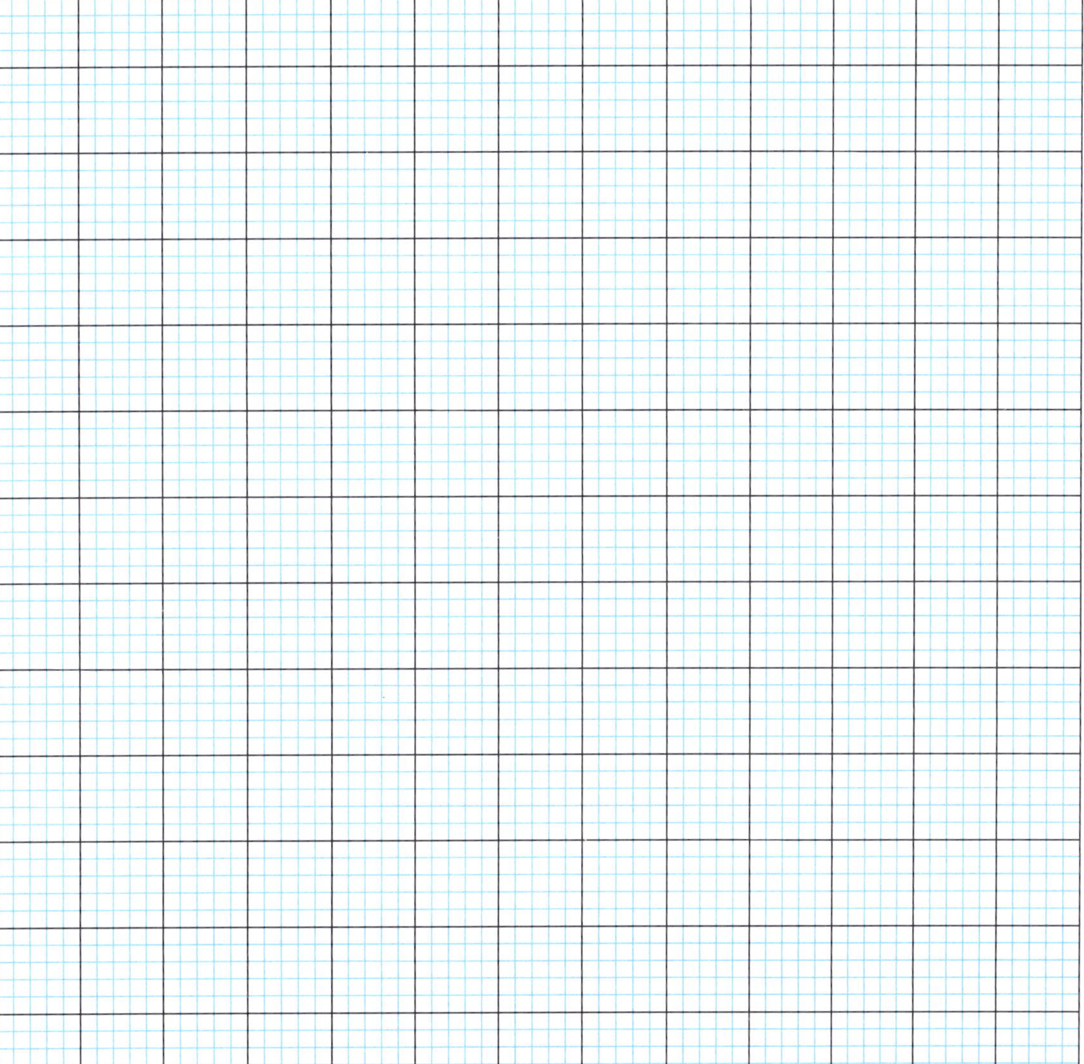

4 With your partner, **investigate** places in the local community that offer dance classes that you might be interested in. List the names of three places, including where they are situated and what type of dance they offer.

investigate to search, inquire into, interpret and draw conclusions about data and information

Venue 1: ______________________________

Venue 2: ______________________________

Venue 3: ______________________________

5 As a class, create a local Dance Directory. In it list places that offer dance classes. Distribute your directory to your class or to others in your school or community.

WORKSHEET 10.2 EXPLORING INDIGENOUS DANCE

Pages 387–8

The Bangarra Dance Theatre is one of Australia's leading Indigenous performing arts organisations.

REKINDLING YOUTH PROGRAM

Weblink 'Rekindling' documentary (10.40min)

Watch the documentary about Bangarra's youth program, Rekindling and reflect on the following questions. You can access the documentary on your Nelson MindTap student book, or google 'Bangarra Rekindling Youth Program'.

1 State what Bangarra Dance Theatre is trying to achieve through its Rekindling Youth Program.

______

______

______

Explain to make an idea or situation plain or clear by describing it in more detail or revealing relevant facts; give an account; provide additional information

2 **Explain** what you think is meant by the term 'custodians of culture'.

______

______

______

3 List three dances that participants learn through the program.

______

______

______

4 Summarise the story that the 'Lost City' aims to communicate.

______

______

______

______

______

5 Explain the significance of visiting 'Lost City'.

______

______

______

9780170465533

6 **Consider** if 'Lost City' is more than just a dance. Why?

Consider to think deliberately or carefully about something, typically before making a decision

AC

7 Many dance movements are used in everyday life and in many sports. In the vision of the dance movements we see participants performing a jeté, or leap. In what other sports is a leap performed?

8 **Define** what is meant by the term 'choreography'.

Define to give the meaning of a word, phrase, or concept

AC

9 **Summarise** the participants' feelings about their performance.

Summarise to give a brief statement of a general theme or major point(s)

AC

10 List three benefits associated with being a part of the Rekindling Youth Program.

Investigation skills

WORKSHEET 10.3 CULTURAL DANCES

Having viewed the footage of Traditional Georgian Dance in your student book, it might inspire you to explore the cultural significance of this dance or one from a different community around the world.

Work with a partner or in a small group to research, investigate and report on a dance from another country or culture.

Choose a cultural community and identify a dance that is performed by this community of people. Use your expert research skills and worksheet questions to help you discover how dance can inform us about cultural identity.

Once you have all your information, create an expert report. This could be in the form of a PowerPoint, Keynote or film documentary in response to the questions. Your teacher will guide you to the format the report should be presented in.

Page 390

1 State the name of the dance.

____

2 From which country or cultural community does the dance originate?

____

Traditional Georgian dance

3 Acknowledge if the dance tells a story and if so elaborate on what this is.

____

____

____

4 Outline why the dance is performed.

____

____

identify to provide an answer from a number of possibilities

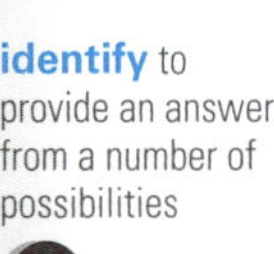

5 **Identify** if the dance is part of a ritual, celebration or ceremony.

____

____

6 Is there a traditional aspect to the dance? If so, explain what this is.

____

____

____

____

____

9780170465533

7 Is the dance still performed in our modern culture?

8 Where is the dance performed?

9 Who can participate in the dance and why?

10 List any special clothing or equipment required to perform the dance.

11 Consider if the dance is performed individually or does it provide an opportunity for social connection.

12 Elaborate on how the dance might provide physical fitness benefits to the participants.

9780170465533

13 Identify any emotional or spiritual significance to the dance. Investigate how participating in the dance could impact the emotional or spiritual wellbeing of the participant.

describe to give an account of characteristics or features

AC

14 **Describe** the overall cultural significance of this dance to the identity, beliefs and values of the people who created it.

9780170465533

WORKSHEET 10.4 CREATING A SPORT DANCE

Pages 390–1

Fill in the dance notation template below as you complete each activity.

Dance notation template	
Group name	**Group members**
	1
	2
	3
	4
Sport focus	
Select a sport or sports your group will focus on	
Coordinator roles	
	1 Steps coordinator
	2 Timing coordinator
	3 Formations coordinator
	4 Presentations coordinator

Step list	
Timing	
How many repetitions of each step?	
Formations	
Use crosses, dots or stick figures to show at least two formations in your dance	
Presentation	
Describe or draw the beginning and ending poses	

9780170465533

WORKSHEET 10.5 REFLECTING AND PROVIDING FEEDBACK BASED ON SET CRITERIA

Pages 401–2

For those groups who do wish to perform their sport dances, it is very important to remember that it shows a lot of courage to get up in front of a group of people and perform. It is important to support and encourage everyone's efforts.

Remember, dancing is about self-expression. Although a group might be performing the same steps, how each individual student interprets and delivers these steps will be unique.

Therefore, if your class chooses to provide feedback to other groups, the feedback must be based only on the set criteria for the dance.

Feedback can only be provided on whether the sport dance:

- has at least four steps (moves) that are repeated twice
- uses a 16-count phrasing for each step (move)
- has at least two formations
- begins and ends with a sporting pose.

HOW TO PROVIDE FEEDBACK

It is always nice to receive encouragement. Remember the following if asked to provide feedback.

1 Begin with a positive feedback statement about how and when the dance met the set criteria.

____

____

____

2 Provide a positive feedback statement about a highlight of the dance.

____

____

____

3 Provide a constructive comment, only if necessary, on how the dance could be improved to meet the set criteria.

____

____

____

Providing feedback related to the set criteria ensures that the dances are critiqued fairly without taking personal dance styles into consideration.

WORKSHEET 10.5 CONTINUED

4 Reflect on your own group's dance. Use the following criteria to assist you to examine and evaluate the process or journey to your creation.

Did your group create a sport dance that:

- had at least four steps (moves) that were repeated twice
- used a 16-count phrasing for each step (move)
- had at least two formations
- began and ended with a sporting pose?

Also include an examination of your coordinator's role, how you demonstrated your leadership skills and your group's collaborative efforts.

Sport dance reflection

Coordinator role reflection

CHAPTER 10 REVIEW

Reflect upon your learning in this chapter. Completing the following sentences will help to structure your thoughts about dance.

1 Information I found interesting was:

2 I was surprised to learn:

3 Information I found useful was:

4 I developed strength in:

5 I am looking forward to:

6 I could change my lifestyle by:

7 Dance and physical expression could benefit others by: